# FRUITING BODIES AND OTHER FUNGI

# Tor Books by Brian Lumley

*Demogorgon*
*The House of Doors*

*THE NECROSCOPE SERIES*
*Necroscope*
*Vamphyri!*
*The Source*
*Deadspeak*
*Deadspawn*

*THE VAMPIRE NOVELS*
*Blood Brothers*
*The Last Aerie (forthcoming)*

*THE PSYCHOMECH SERIES*
*Psychomech*
*Psychosphere*
*Psychamok*

# FRUITING BODIES AND OTHER FUNGI

## BRIAN LUMLEY

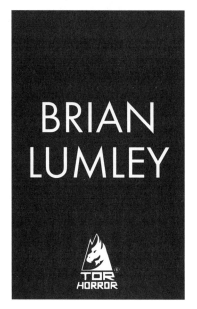

TOR
HORROR

A TOM DOHERTY ASSOCIATES BOOK
NEW YORK

FRUITING BODIES AND OTHER FUNGI

Copyright © 1993 by Brian Lumley

This book is printed on acid-free paper.

A Tor Book
Published by Tom Doherty Associates, Inc.
175 Fifth Avenue
New York, N.Y. 10010

Tor® is a registered trademark of Tom Doherty Associates, Inc.

Library of Congress Cataloging-in-Publication Data

Lumley, Brian.
    Fruiting bodies and other fungi / Brian Lumley.
        p.  cm.
    "A Tom Doherty Associates book."
    ISBN 0-312-85458-7
    I. Horror tales, English.  I. Title.
  PR6062.U45F78   1993
  823'.914—dc20                      92-42574
                                         CIP

First edition: February 1993

Printed in the United States of America

0 9 8 7 6 5 4 3 2

# Copyright Information

*For Alex and Kate*

# Contents

# Introduction

O nce upon a time, things used to go *bump, bump, bump* in the night. Now they go *gloop, spurt,* and very often *splatter!* I believe it's largely a fad, where these so-called "new breed" writers are vying with each other to see how far they can push it before even *they* throw up! A sort of literary (?) chicken race—to the bathroom.

Except . . . it suddenly dawns on me that what I've written above might give the wrong impression. It might appear that I am dead against splatterpunk, and I'm not. I'm not dead against *any* sort of writer—male, female, or undecided—or any sort of writing . . . well, with the possible exception of what I term "pretentious pap," which even so I wouldn't attempt to ban, censor, or suppress in any way, shape, or form. The Great Reading Public will decide that, just as it has made similar decisions in the past, all in its own sweet time. But I do know what I like, and what I will or won't read, watch on TV, listen to, eat, etc.

What I mean is, I can't eat tripe not because it's inedible (though it certainly is to me!) but because it turns my stomach. And I can't watch Australian soap operas for the same reason. And the neutered voices of certain young British pop stars hurt my ears even worse than Tiny Tim's cacophonies did in their time. But all of these things have their aficionados, and no way I'm going to deny anyone access to his or her own taste buds.

Quite a few writers appear to wish that the Good Old Days of horror were back again. I remember advising one such not to worry about it. "You die if you worry and you die if you don't," I said. "So why worry? Whether you love or hate it, splatter is, it exists, it's here . . . though not necessarily to stay. For like everything else in the arts—especially in the pop arts, and specifically in what should be the literary art of 'entertainment by frisson'—the good stuff will swim and the bad stuff will sink into oblivion. So give it a year or two and see where it's at. Come the Year 2000 and present-day punks who can actually hack it (?) will be writing, and those who can't will still be splattering. Except splatter may not seem too much fun anymore (picture the annual World Splattercon, with the awards panel handing out statuettes in the form of exquisitely sculpted lower intestines or working models of heaving stomachs) and the splatterers may not be pulling too many readers. For if books don't sell, publishers don't publish, and it's still a basic requirement of stories and novels that they are written. That's *written,* not vomited or ejaculated, or ejected from some nameless nether orifice, but written!" I said something like that, anyway, to my writer friend.

But . . . no sooner had I got that lot off my chest than another friend (oh, yes, I still have a few) surprised me by suggesting: "But aren't you just a teensy weensy bit splattery yourself? I mean, that opening scene in *Necroscope,* where Dragosani eviscerates the dead Russian to tear his secrets right out of his guts, is sort of . . . well, it's—"

Splatter?

I couldn't see it. I still can't. Dragosani's violence was integral, intrinsic, *necessary!* What would the book be without it? And now I'm starting to sound like a young starlet apologizing for a certain notorious sex scene, which she swears was all done "in the best *possible* taste!" Nevertheless, I'll take a chance and likewise swear that Dragosani's violence wasn't "gratuitous," that I was just telling the story. In fact I'll cop right out by adding: "Anyway, how in hell can you hold *me* responsible for what Dragosani did, eh?"

So perhaps, like love or beauty, splatter is all in the eye of the beholder. And maybe not all that much has changed anyway, since the Good Old Days of horror. I remember a quote I saw once on the back of a rather excellent paperback: "Raw Head," (?) "and Bloody Bones, and Tales Filled with Graveyard Stenches!" I may have it a little wrong, but close enough. And that was thirty years ago on the back of a collection of Lovecraft's stories! It's a sure bet that if the splatterpunks had been around then, they'd have bought that one.

Today . . . well, here in Lovecraft's centenary year, while the peculiar Old Gent of Providence may be long gone, what he left behind is still very much in evidence. Like old tentacle-wreathed Cthulhu himself, HPL seems eternal, not really dead. Certainly his work is still alive and kicking, though it has to be admitted that it's starting to labor a bit under the weight of years. But as for the work of a good many of his contemporaries in *Weird Tales* and such, where is it now? Mostly thrown out, that's where. All but forgotten. Sunk like R'lyeh. And it won't come back up.

As then, so now. The writers of works that were worth remembering and revisiting are revered and reprinted. Those who aren't won't be. The cream floats, and the dregs go spiraling down. . . .

As I process this up onto my screen, I've been in the writing business twenty-two years, probably twenty-four by the time it sees print. But I've been a professional, a full-timer, for little more than a third of that time. In the last eight years I've seen a lot of the subtlety go right out of the horror story; not all of it, but a lot. And I find myself at a loss to explain it. Are writers any less capable? Readers any less demanding? *Have* our senses been so thoroughly blunted by the battering they've taken from all this "gratuitous violence" stuff?

Me, I'm someone who believes that you shouldn't knock it till you've tried it—literarily speaking, you understand. Splatterpunk? I've tried it, and freely admit that I can't. The one or two pieces I finished were unsatisfactory, lacking in one way or another, so that even though I wrote them, still I don't especially like them. Instead of that old black magic, that frisson, that shudder of delight, they gave me the clammy feel of something else. I may have been horrified, but I was *not* entertained.

*Whoa!* I think I may have hit upon the answer. Maybe it's as simple as this: horror stories are still horror stories, but in too many cases the function of the modern variety is no longer to entertain . . . *just* to horrify!

In *Fruiting Bodies* I would like to think I'm taking you back, oh, at least part of the way to those Good Old Days of *bump, bump, bump* in the night; and I'm sorry, but what you had for supper is strictly between you and your tripes. You won't be parting with it on my behalf. Here you'll find stories from across the full span of my twenty-odd years in the business, from my first book right up to the present day and some of my most recent writing, but I don't think you'll find a one to throw up by.

On the other hand, if I *should* spill a little more than a drop or two, then I beg your forgiveness that I forgot my prime objective, which was to entertain, and I hope you'll believe me when I say that to the best of my knowledge and intentions it was all done "in the best *possible* taste!"

Brian Lumley
March 1990

# Fruiting Bodies

*This first story won a British Fantasy Award in 1989. It had some stiff competition, and I count myself lucky to have won. Whether it's frightening or not is for you to decide. If it's entertaining and gives that certain frisson, then I'm satisfied. One thing's for sure: there isn't any blood here. Mushrooms don't bleed.*

My great-grandparents, and my grandparents after them, had been Easingham people; in all likelihood my parents would have been, too, but the old village had been falling into the sea for three hundred years and hadn't much looked like stopping, and so I was born in Durham City instead. My grandparents, both sets, had been among the last of the village people to move out, buying new homes out of a government-funded disaster grant. Since when, as a kid, I had been back to Easingham only once.

My father had taken me there one spring when the tides were high. I remember how there was still some black, crusty snow lying in odd corners of the fields, colored by soot and smoke, as all things were in those days in the Northeast. We'd gone to Easingham because the unusually high tides had been at it again, chewing away at the shale cliffs, reducing shoreline and derelict village both as the North Sea's breakers crashed again and again on the shuddering land.

And of course we had hoped (as had the two hundred or so other sightseers gathered there that day) to see a house or two go down in smoking ruin, into the sea and the foaming spray. We witnessed no such spectacle; after an hour, cold and wet from the salt moisture in the air, we piled back into the family car and returned to Durham. Easingham's main street, or what had once been the main street, was teetering on the brink as we left. But by nightfall that street was no more. We'd missed it: a further twenty feet of coastline, a bite one street deep and a few yards more than one street long, had been undermined, toppled, and gobbled up by the sea.

That had been that. Bit by bit, in the quarter century between then and now, the rest of Easingham had also succumbed. Now only a house or two remained—no more than a handful in all—and all falling into decay, while the closest lived-in buildings were those of a farm all of a mile inland from the cliffs. Oh, and of course there was one other inhabitant: old Garth Bentham, who'd been demolishing the old houses by hand and selling bricks and timbers from the village for years. But I'll get to him shortly.

So there I was last summer, back in the Northeast again, and when my business was done of course I dropped in and stayed overnight with the Old Folks at their Durham cottage. Once a year at least I made a point of seeing them, but last year in particular I noticed how time was creeping up on them. The "Old Folks"; well, now I saw that they really were old, and I determined that I must start to see a lot more of them.

Later, starting in on my long drive back down to London, I remembered that time when the Old Man had taken me to Easingham to see the houses tottering on the cliffs.

And probably because the place was on my mind, I inadvertently turned off my route and in a little while found myself heading for the coast. I could have turned round right there and then—indeed, I intended to do so—but I'd got to wondering about Easingham and how little would be left of it now, and before I knew it . . .

Once I'd made up my mind, Middlesborough was soon behind me, then Guisborough, and in no time at all I was on the old road to the village. There had only ever been one way in and out, and this was it: a narrow road, its surface starting to crack now, with tall hedgerows broken here and there, letting you look through to where fields rolled down to the cliffs. A beautiful day, with sea gulls wheeling overhead, a salt tang coming in through the wound-down windows, and a blue sky coming down to merge with . . . with the blue-grey of the North Sea itself! For cresting a rise, suddenly I was there.

An old, leaning wooden signpost said EASINGH—, for the tail had been broken off or rotted away, and "the village" lay at the end of the road. But right there, blocking the way, a metal barrier was set in massive concrete posts and carried a sign bearing the following warning:

DANGER!
SEVERE CLIFF SUBSIDENCE.
NO VEHICLES BEYOND THIS POINT . . .

I turned off the car's motor, got out, leaned on the barrier. Before me the road went on—and disappeared only thirty yards ahead. And there stretched the new rim of the cliffs. Of the village, Easingham itself—forget it! On this side of the cliffs, reaching back on both sides of the road behind overgrown gardens, weedy paths, and driveways, here stood the empty shells of what had once been residences of the "posh" folks of Easingham. Now, even on a day as lovely as this one, they were morose in their desolation.

The windows of these derelicts, where there were windows, seemed to gaze gauntly down on approaching doom, like old men in twin rows of deathbeds. Brambles and ivy were rank; the whole place seemed despairing as the cries of the gulls rising on the warm air; Easingham was a place no more.

Not that there had ever been a lot of it. Three streets lengthwise with a few shops; two more, shorter streets cutting through the three at right angles and going down to the cliffs and the vertiginous wooden steps that used to climb down to the beach, the bay, the old harbor, and fish market; and standing over the bay, a Methodist church on a jutting promontory, which in the old times had also served as a lighthouse. But now—

No streets, no promontory or church, no harbor, fish market, rickety steps. No Easingham.

"Gone, all of it," said a wheezy, tired old voice from directly behind me, causing me to start. "Gone forever, to the devil and the deep blue sea!"

I turned, formed words, said something barely coherent to the leathery old scarecrow of a man I found standing there.

"Eh? Eh?" he said. "Did I startle you? I have to say you startled me! First car I've seen in a three-month! After bricks, are you? Cheap bricks? Timber?"

"No, no," I told him, finding my voice. "I'm—well, sight-seeing, I suppose." I shrugged. "I just came to see how the old village was getting on. I didn't live here, but a long line of my people did. I just thought I'd like to see how much was left—while it *was* left! Except it seems I'm too late."

"Oh, aye, too late." He nodded. "Three or four years too late. That was when the last of the old fishing houses went down: four years ago. Sea took 'em. Takes six or seven feet of cliff every year. Aye, and if I lived long enough it would take me, too. But it won't 'cos I'm getting on a bit." And he grinned and nodded, as if to say: So that's that! "Well, well, sight-seeing! Not much to see, though, not now. Do you fancy a coffee?"

Before I could answer he put his fingers to his mouth and blew a piercing whistle, then paused and waited, shook his head in puzzlement. "Ben," he explained. "My old dog. He's not been himself lately and I don't like him to stray too far. He was out all night, was Ben. Still, it's summer, and there may have been a bitch about. . . ."

While he had talked I'd looked him over and decided that I liked him. He reminded me of my own grandfather, what little I could remember of him. Grandad had been a miner in one of the colliery villages farther north, retiring here to doze and dry up and die—only to find himself denied the choice. The sea's incursion had put paid to that when it finally made the place untenable. I fancied this old lad had been a miner, too. Certainly he bore the scars, the stigmata, of the miner: the dark, leathery skin with black specks bedded in; the bad, bowed legs; the shortness of breath, making for short sentences. A generally gritty appearance overall, though I'd no doubt he was clean as fresh scrubbed.

"Coffee would be fine," I told him, holding out my hand. "Greg's my name—Greg Lane."

He took my hand, shook it warmly, and nodded. "Garth Bentham," he said. And then he set off stiffly back up the crumbling road some two or three houses, turning right into an overgrown garden through a fancy wooden gate recently painted white. "I'd intended doing the whole place up," he said, as I followed close behind. "Did the gate, part of the fence, ran out of paint!"

Before letting us into the dim interior of the house, he paused and whistled again for Ben, then worriedly shook his head in something of concern. "After rats in the old timber yard again, I suppose. But God knows I wish he'd stay out of there!"

Then we were inside the tiny cloakroom, where the sun filtered through fly-specked windows and probed golden searchlights on a few fairly dilapidated furnishings and the brassy face of an old grandfather clock that clucked like a mechanical hen. Dust motes drifted like tiny planets in a cosmos of faery, eddying round my host where he guided me through a door and into his living room. Where the dust had settled on the occasional ledge, I noticed that it was tinged red, like rust.

"I cleaned the windows in here," Garth informed, "so's to see the sea. I like to know what it's up to!"

"Making sure it won't creep up on you." I nodded.

His eyes twinkled. "Nah, just joking," he said, tapping on the side of his blue-veined nose. "No, it'll be ten or even twenty years before all this goes, but I don't have that long. Five if I'm lucky. I'm sixty-eight, after all!"

Sixty-eight! Was that really to be as old as all that? But he was probably right: a lot of old-timers from the mines didn't even last *that* long, not entirely mobile and coherent, anyway. "Retiring at sixty-five doesn't leave a lot, does it?" I said. "Of time, I mean."

He went into his kitchen, called back: "Me, I've been here a ten-year. Didn't retire, quit! Stuff your pension, I told 'em. I'd rather have my lungs, what's left of 'em. So I came here, got this place for a song, take care of myself and my old dog, and no one to tip my hat to and no one to bother me. I get a letter once a fortnight from my sister in Dunbar, and one of these days the postman will find me stretched out in here and he'll think: 'Well, I needn't come out here anymore.'"

He wasn't bemoaning his fate, but I felt sorry for him anyway. I settled myself on a dusty settee, looked out of the window down across his garden of brambles to the sea's horizon. A great curved millpond—for the time being. "Didn't you have any savings?" I could have bitten my tongue off the moment I'd said it, for that was to imply he hadn't done very well for himself.

Cups rattled in the kitchen. "Savings? Lad, when I was a young 'un I had three things: my lamp, my helmet, and a pack of cards. If it wasn't pitch-'n-toss with weighted pennies on the beach banks, it was three-card brag in the back room of the pub. Oh, I was a game gambler, right enough, but a bad one. In my blood, like my Old Man before me. My mother never did see a penny; nor did my wife, I'm ashamed to say, before we moved out here—God bless her! Savings? That's a laugh. But out here there's no bookie's runner, and you'd be damned hard put to find a card school in Easingham these days! What the hell," he shrugged as he stuck his head back into the room, "it was a life. . . ."

We sipped our coffee. After a while I said, "Have you been on your own very long? I mean . . . your wife?"

"Lily-Anne?" He glanced at me, blinked, and suddenly there was a peculiar expression on his face. "On my own, you say. . . ." He straightened his shoulders, took a deep breath. "Well, I *am* on my own in a way, and in a way I'm not. I have Ben—or would have if he'd get done with what he's doing and come home—and Lily-Anne's not all that far away. In fact, sometimes I suspect she's sort of watching over me, keeping me company, so to speak. You know, when I'm feeling especially lonely."

"Oh?"

"Well." He shrugged again. "I mean she *is* here, now isn't she." It was a statement, not a question.

"Here?" I was starting to have my doubts about Garth Bentham.

"I had her buried here." He nodded, which explained what he'd said and produced a certain sensation of relief in me. "There was a Methodist church here once over, with its own burying ground. The church went a donkey's years ago, of course, but the old graveyard was still here when Lily-Anne died."

"Was?" Our conversation was getting one-sided.

"Well, it still is—but right on the edge, so to speak. It wasn't so bad then, though, and so I got permission to have a service done here, and down she went where I could go and see her. I still do go to see her, of course, now and then. But in another year or two . . . the sea . . ." He shrugged again. "Time and the tides, they wait for no man."

We finished our coffee. I was going to have to be on my way soon, and suddenly I didn't like the idea of leaving him. Already I could feel the loneliness creeping in. Perhaps he sensed my restlessness or something. Certainly I could see that he didn't want me to go just yet. In any case, he said:

"Maybe you'd like to walk down with me past the old timber yard, visit her grave. Oh, it's safe enough, you don't have to worry. We may even come across old Ben down there. He sometimes visits her, too."

"Ah, well I'm not too sure about that," I answered. "The time, you know?" But by the time we got down the path to the gate I was asking: "How far is the churchyard, anyway?" Who could tell, maybe I'd find some long-lost Lanes in there! "Are there any old markers left standing?"

Garth chuckled and took my elbow. "It makes a change to have some company," he said. "Come on, it's this way."

He led the way back to the barrier where it spanned the road, bent his back, and ducked groaning under it, then turned left up an overgrown communal path between gardens where the houses had been stepped down the declining gradient. The detached bungalow on our right—one of a pair still standing, while a third slumped on the raw edge of oblivion—had decayed almost to the point where it was collapsing inward. Brambles luxuriated everywhere in its garden, completely enclosing it. The roof sagged and a chimney threatened to topple, making the whole structure seem highly suspect and more than a little dangerous.

"Partly subsidence, because of the undercutting action of the sea," Garth explained, "but mainly the rot. There was a lot of wood in these places, but it's all being eaten away. I made myself a living, barely, out of the old bricks and timber in Easingham, but now I have to be careful. Doesn't do to sell stuff with the rot in it."

"The rot?"

He paused for breath, leaned a hand on one hip, nodded and frowned. "Dry rot," he said. "Or *Merulius lacrymans* as they call it in the books. It's been bad these last three years. Very bad! But when the last of these old houses are gone, and what's left of the timber yard, then it'll be gone, too."

"It?" We were getting back to single-word questions again. "The dry rot, you mean? I'm afraid I don't know very much about it."

"Places on the coast are prone to it," he told me. "Whitby, Scarborough, places like that. All the damp sea spray and the bad plumbing, the rains that come in and the inadequate drainage. That's how it starts. It's a fungus, needs a lot of moisture—to get started, anyway. You don't know much about it? Heck, I used to think I knew *quite* a bit about it, but now I'm not so sure!"

By then I'd remembered something. "A friend of mine in London did mention to me how he was having to have his flat treated for it," I said, a little lamely. "Expensive, apparently."

Garth nodded, straightened up. "Hard to kill," he said. "And when it's active, moves like the plague! It's active here, now! Too late for Easingham, and who gives a damn anyway? But you tell that friend of yours to sort out his exterior maintenance first: the guttering and the drainage.

Get rid of the water spillage, then deal with the rot. If a place is dry and airy, it's OK. Damp and musty spells danger!"

I nodded. "Thanks, I will tell him."

"Want to see something?" said Garth. "I'll show you what old *Merulius* can do. See here, these old paving flags? See if you can lever one up a bit." I found a piece of rusting iron stave and dragged it out of the ground where it supported a rotting fence, then forced the sharp end into a crack between the overgrown flags. And while I worked to loosen the paving stone, old Garth stood watching and carried on talking.

"Actually, there's a story attached, if you care to hear it," he said. "Probably all coincidental or circumstantial, or some other big word like that—but queer the way it came about all the same."

He was losing me again. I paused in my levering to look bemused (and maybe to wonder what on Earth I was doing here), then grunted, and sweated, gave one more heave, and flipped the flag over onto its back. Underneath was hard-packed sand. I looked at it, shrugged, looked at Garth.

He nodded in that way of his, grinned, said: "Look. Now tell me what you make of this!"

He got down on one knee, scooped a little of the sand away. Just under the surface his hands met some soft obstruction. Garth wrinkled his nose and grimaced, got his face down close to the earth, blew until his weakened lungs started him coughing. Then he sat back and rested. Where he'd scraped and blown the sand away, I made out what appeared to be a grey fibrous mass running at right angles right under the pathway. It was maybe six inches thick, looked like tightly packed cotton wool. It might easily have been glass fiber lagging for some pipe or other, and I said as much.

"But it isn't," Garth contradicted me. "It's a root, a

feeler, a tentacle. It's old man cancer himself—timber cancer—on the move and looking for a new victim. Oh, you won't see him moving," that strange look was back on his face, "or at least you shouldn't—but he's at it anyway. He finished those houses there," he nodded at the derelicts stepping down toward the new cliffs, "and now he's gone into this one on the left here. Another couple of summers like this 'un and he'll be through the entire row to my place. Except maybe I'll burn him out first."

"You mean this stuff—this fiber—is dry rot?" I said. I stuck my hand into the stuff and tore a clump out. It made a soft tearing sound, like damp chipboard, except it was dry as old paper. "How do you mean, you'll 'burn him out'?"

"I mean like I say," said Garth. "I'll search out and dig up all these threads—mycelium, they're called—and set fire to 'em. They smoulder right through to a fine white ash. And God—it *stinks!* Then I'll look for the fruiting bodies, and—"

"The what?" His words had conjured up something vaguely obscene in my mind. "Fruiting bodies?"

"Lord, yes!" he said. "You want to see? Just follow me."

Leaving the path, he stepped over a low brick wall to struggle through the undergrowth of the garden on our left. Taking care not to get tangled up in the brambles, I followed him. The house seemed pretty much intact, but a bay window in the ground floor had been broken and all the glass tapped out of the frame. "My winter preparations," Garth explained. "I burn wood, see? So before winter comes, I get into a house like this one, rip out all the wooden fixings and break 'em down ready for burning. The wood just stays where I stack it, all prepared and waiting for the bad weather to come in. I knocked this window out last week, but I've not been inside yet. I could smell it, see?" He tapped his nose. "And I didn't much care for all those spores on my lungs."

He stepped up on a pile of bricks, got one leg over the sill, and stuck his head inside. Then, turning his head in all directions, he systematically sniffed the air. Finally he seemed satisfied and disappeared inside. I followed him. "Spores?" I said. "What sort of spores?"

He looked at me, wiped his hand along the window ledge, held it up so that I could see the red dust accumulated on his fingers and palm. *"These* spores," he said. "Dry-rot spores, of course! Haven't you been listening?"

"I *have* been listening, yes," I answered sharply. "But I ask you: spores, mycelium, fruiting bodies? I mean, I thought dry rot was just, well, rotting wood!"

"It's a fungus," he told me, a little impatiently. "Like a mushroom, and it spreads in much the same way. Except it's destructive, and once it gets started it's bloody hard to stop!"

"And you, an ex–coal miner," I stared at him in the gloom of the house we'd invaded, "you're an expert on it, right? How come, Garth?"

Again there was that troubled expression on his face, and in the dim interior of the house he didn't try too hard to mask it. Maybe it had something to do with that story he'd promised to tell me, but doubtless he'd be as circuitous about that as he seemed to be about everything else. "Because I've read it up in books, that's how," he finally broke into my thoughts. "To occupy my time. When it first started to spread out of the old timber yard, I looked it up. It's—" He gave a sort of grimace. "—it's *interesting,* that's all."

By now I was wishing I was on my way again. But by that I mustn't be misunderstood: I'm an able-bodied man and I wasn't afraid of anything—and certainly not of Garth himself, who was just a lonely, canny old-timer—but all of this really was getting to be a waste of my time. I had just made my mind up to go back out through the window when he caught my arm.

"Oh, *yes!*" he said. "This place is really ripe with it! Can't you smell it? Even with the window bust wide open like this, and the place nicely dried out in the summer heat, still it's stinking the place out. Now just you come over here and you'll see what you'll see."

Despite myself, I was interested. And indeed I could smell . . . something. A cloying mustiness? A mushroomy taint? But not the nutty smell of fresh field mushrooms. More a sort of vile stagnation. Something dead might smell like this, long after the actual corruption has ceased. . . .

Our eyes had grown somewhat accustomed to the gloom. We looked about the room. "Careful how you go," said Garth. "See the spores there? Try not to stir them up too much. They're worse than snuff, believe me!" He was right: the red dust lay fairly thick on just about everything. By "everything" I mean a few old sticks of furniture, the worn carpet under our feet, the skirting-board, and various shelves and ledges. Whichever family had moved out of here, they hadn't left a deal of stuff behind them.

The skirting was of the heavy, old-fashioned variety: an inch and a half thick, nine inches deep, with a fancy moulding along the top edge; they hadn't spared the wood in those days. Garth peered suspiciously at the skirting-board, followed it away from the bay window, and paused every pace to scrape the toe of his boot down its face. And eventually when he did this—suddenly the board crumbled to dust under the pressure of his toe!

It was literally as dramatic as that: the white paint cracked away and the timber underneath fell into a heap of black, smoking dust. Another pace and Garth kicked again, with the same result. He quickly exposed a ten-foot length of naked wall, on which even the plaster was loose and flaky, and showed me where strands of the cotton-wool mycelium had come up between the brickwork and the plaster from below. "It sucks the cellulose right out of wood," he said. "Gets right into brickwork, too. Now look here," and he pointed at the old carpet under his feet. The threadbare weave showed a sort of raised floral blossom or stain, like a blotch or blister, spreading outward away from the wall.

Garth got down on his hands and knees. "Just look at this," he said. He tore up the carpet and carefully laid it back. Underneath, the floorboards were warped, dark stained, shriveled so as to leave wide gaps between them. And up through the gaps came those white, etiolated threads, spreading themselves along the underside of the carpet.

I wrinkled my nose in disgust. "It's like a disease," I said.

"It *is* a disease!" he corrected me. "It's a cancer, and houses die of it!" Then he inhaled noisily, pulled a face of his own, and said: "Here. Right here." He pointed at the warped, rotting floorboards. "The very heart of it. Give me a hand." He got his fingers down between a pair of boards and gave a tug, and it was at once apparent that he wouldn't be needing any help from me. What had once been a stout wooden floorboard a full inch thick was now brittle as dry bark. It cracked upward, flew apart, revealed the dark cavities between the floor joists. Garth tossed bits of crumbling wood aside, tore up more boards; and at last "the very heart of it" lay open to our inspection.

"There!" said Garth with a sort of grim satisfaction. He stood back and wiped his hands down his trousers. "Now *that* is what you call a fruiting body!"

It was roughly the size of a football, if not exactly that shape. Suspended between two joists in a cradle of fibers, and adhering to one of the joists as if partly flattened to it, the thing might have been a great, too-ripe tomato. It was bright yellow at its center, banded in various shades of yellow from the middle out. It looked freakishly weird, like

a bad joke: this lump of . . . of *stuff*—never a mushroom—
just nestling there between the joists.

Garth touched my arm and I jumped a foot. He said:
"You want to know where all the moisture goes—out of
this wood, I mean? Well, just touch it."

"Touch . . . that?"

"Heck, it can't bite you! It's just a fungus."

"All the same, I'd rather not," I told him.

He took up a piece of floorboard and prodded the
thing—and it squelched. The splintered point of the wood
sank into it like jelly. Its heart was mainly liquid, porous as
a sponge. "Like a huge egg yolk, isn't it?" he said, his voice
very quiet. He was plainly fascinated.

Suddenly I felt nauseous. The heat, the oppressive
closeness of the room, the spore-laden air. I stepped dizzily backward and stumbled against an old armchair. The
rot had been there, too, for the chair just fragmented into
a dozen pieces that puffed red dust all over the place. My
foot sank right down through the carpet and mushy
boards into darkness and stench—and in another moment
I'd panicked.

Somehow I tumbled myself back out through the window and ended up on my back in the brambles. Then Garth
was standing over me, shaking his head and tut-tutting.
"Told you not to stir up the dust," he said. "It chokes your
air and stifles you. Worse than being down a pit. Are you
all right?"

My heart stopped hammering and I was, of course, all
right. I got up. "A touch of claustrophobia," I told him. "I
suffer from it at times. Anyway, I think I've taken up
enough of your time, Garth. I should be getting on my
way."

"What?" he protested. "A lovely day like this and you
want to be driving off somewhere? And besides, there were
things I wanted to tell you, and others I'd ask you—and we

haven't been down to Lily-Anne's grave." He looked disappointed. "Anyway, you shouldn't be driving if you're feeling all shaken up. . . ."

He was right about that part of it, anyway: I did feel shaky, not to mention foolish! And perhaps more importantly, I was still very much aware of the old man's loneliness. What if it was my mother who'd died, and my father had been left on his own up in Durham? "Very well," I said, at the same time damning myself for a weak fool, "let's go and see Lily-Anne's grave."

"Good!" Garth slapped my back. "And no more diversions—we go straight there."

Following the paved path as before and climbing a gentle rise, we started walking. We angled a little inland from the unseen cliffs where the green, rolling fields came to an abrupt end and fell down into the sea; and as we went I gave a little thought to the chain of incidents in which I'd found myself involved through the last hour or so.

Now, I'd be a liar if I said that nothing had struck me as strange in Easingham, for quite a bit had. Not least the dry rot: its apparent profusion and migration through the place, and old Garth's peculiar knowledge and understanding of the stuff. His—affinity?—with it. "You said there was a story attached," I reminded him. ". . . To that horrible fungus, I mean."

He looked at me sideways, and I sensed he was on the point of telling me something. But at that moment we crested the rise and the view just took my breath away. We could see for miles up and down the coast: to the slow, white breakers rolling in on some beach way to the north, and southward to a distance-misted seaside town that might even be Whitby. And we paused to fill our lungs with good air blowing fresh off the sea.

"There," said Garth. "And how's this for freedom? Just me and old Ben and the gulls for miles and miles, and I'm not so sure but that this is the way I like it. Now wasn't it worth it to come up here? All this open space and the great curve of the horizon . . ." Then the look of satisfaction slipped from his face to be replaced by a more serious expression. "There's old Easingham's cemetery—what's left of it."

He pointed down toward the cliffs, where a badly weathered stone wall formed part of a square whose sides would have been maybe fifty yards long in the old days. But in those days there'd also been a stubby promontory and a church. Now only one wall, running parallel with the path, stood complete—beyond which two-thirds of the churchyard had been claimed by the sea. Its occupants, too, I supposed.

"See that half-timbered shack," said Garth, pointing, "at this end of the cemetery? That's what's left of Johnson's Mill. Johnson's sawmill, that is. That shack used to be Old Man Johnson's office. A long line of Johnsons ran a couple of farms that enclosed all the fields round here right down to the cliffs. Pasture, mostly, with lots of fine animals grazing right here. But as the fields got eaten away and the buildings themselves started to be threatened, that's when half the Johnsons moved out and the rest bought a big house in the village. They gave up farming and started the mill, working timber for the local building trade. . . .

"Folks round here said it was a sin, all that noise of sawing and planing, right next door to a churchyard. But . . . it was Old Man Johnson's land after all. Well, the sawmill business kept going till a time some seven years ago, when a really bad blow took a huge bite right out of the bay one night. The seaward wall of the graveyard went, and half of the timber yard, too, and that closed old Johnson down. He sold what machinery he had left, plus a few stacks of good oak that hadn't suffered, and moved out lock, stock, and barrel. Just as well, for the very next spring his big house and two others close to the edge of the cliffs got taken. The sea gets 'em all in the end.

"Before then, though—at a time when just about everybody else was moving out of Easingham—Lily-Anne and me had moved in! As I told you, we got our bungalow for a song, and of course we picked ourselves a house standing well back from the brink. We were getting on a bit; another twenty years or so should see us out; after that the sea could do its worst. But . . . well, it didn't quite work out that way."

While he talked, Garth had led the way down across the open fields to the graveyard wall. The breeze was blustery here and fluttered his words back into my face:

"So you see, within just a couple of years of our settling here, the village was derelict, and all that remained of people was us and a handful of Johnsons still working the mill. Then Lily-Anne came down with something and died, and I had her put down in the ground here in Easingham—so's I'd be near her, you know?

"That's where the coincidences start to come in, for she went only a couple of months after the shipwreck. Now I don't suppose you'd remember that; it wasn't much, just an old Portuguese freighter that foundered in a storm. Lifeboats took the crew off, and she'd already unloaded her cargo somewhere up the coast, so the incident didn't create much of a to-do in the newspapers. But she'd carried a fair bit of hardwood ballast, that old ship, and balks of the stuff would keep drifting ashore: great long twelve-by-twelves of it. Of course, Old Man Johnson wasn't one to miss out on a bit of good timber like that, not when it was being washed up right on his doorstep, so to speak. . . .

"Anyway, when Lily-Anne died I made the proper arrangements, and I went down to see old Johnson who told me he'd make me a coffin out of this Haitian hardwood."

"Haitian?" Maybe my voice showed something of my surprise.

"That's right," said Garth, more slowly. He looked at me wonderingly. "Anything wrong with that?"

I shrugged, shook my head. "Rather romantic, I thought," I said. "Timber from a tropical isle."

"I thought so, too," he agreed. And after a while he continued: "Well, despite having been in the sea, the stuff could still be cut into fine, heavy panels, and it still French-polished to a beautiful finish. So that was that: Lily-Anne got a lovely coffin. Except—"

"Yes?" I prompted him.

He pursed his lips. "Except I got to thinking—later, you know—as to how maybe the rot came here in that wood. God knows it's a damn funny variety of fungus after all. But then this Haiti—well, apparently it's a damned funny place. They call it 'the Voodoo Island,' you know?"

"Black magic?" I smiled. "I think we've advanced a bit beyond thinking such as that, Garth."

"Maybe and maybe not," he answered. "But voodoo or no voodoo, it's still a funny place, that Haiti. Far away and exotic . . ."

By now we'd found a gap in the old stone wall and climbed over the tumbled stones into the graveyard proper. From where we stood, another twenty paces would take us right to the raw edge of the cliff where it sheared dead straight through the overgrown, badly neglected plots and headstones. "So here it is," said Garth, pointing. "Lily-Anne's grave, secure for now in what little is left of Easingham's old cemetery." His voice fell a little, grew ragged: "But you know, the fact is I wish I'd never put her down here in the first place. And I'd give anything that I hadn't buried her in that coffin built of Old Man Johnson's ballast wood."

The plot was a neat oblong picked out in oval pebbles. It

had been weeded round its border, and from its bottom edge to the foot of the simple headstone it was decked in flowers, some wild and others cut from Easingham's deserted gardens. It was deep in flowers, and the ones underneath were withered and had been compressed by those on top. Obviously Garth came here more often than just "now and then." It was the only plot in sight that had been paid any sort of attention, but in the circumstances that wasn't surprising.

"You're wondering why there are so many flowers, eh?" Garth sat down on a raised slab close by.

I shook my head, sat down beside him. "No, I know why. You must have thought the world of her."

"You don't know why," he answered. "I did think the world of her, but that's not why. It's not the only reason, anyway. I'll show you."

He got down on his knees beside the grave, began laying aside the flowers. Right down to the marble chips he went, then scooped an amount of the polished gravel to one side. He made a small mound of it. Whatever I had expected to see in the small excavation, it wasn't the cylindrical, fibrous surface—like the upper section of a lagged pipe—that came into view. I sucked in my breath sharply.

There were tears in Garth's eyes as he flattened the marble chips back into place. "The flowers are so I won't see it if it ever breaks the surface," he said. "See, I can't bear the thought of that filthy stuff in her coffin. I mean, what if it's like what you saw under the floorboards in that house back there?" He sat down again, and his hands trembled as he took out an old wallet and removed a photograph to give it to me. "That's Lily-Anne," he said. "But God!—I don't like the idea of that stuff fruiting on her. . . ."

Aghast at the thoughts his words conjured, I looked at the photograph. A homely woman in her late fifties, seated in a chair beside a fence in a garden I recognized as Garth's.

Except the garden had been well-tended then. One shoulder seemed slumped a little, and though she smiled, still I could sense the pain in her face. "Just a few weeks before she died," said Garth. "It was her lungs. Funny that I worked in the pit all those years, and it was her lungs gave out. And now she's here, and so's this stuff."

I had to say something. "But . . . where did it come from. I mean, how did it come, well, here? I don't know much about dry rot, no, but I would have thought it confined itself to houses."

"That's what I was telling you," he said, taking back the photograph. "The British variety does. But not this stuff. It's weird and different! That's why I think it might have come here with that ballast wood. As to how it got into the churchyard: that's easy. Come and see for yourself."

I followed him where he made his way between the weedy plots toward the leaning, half-timbered shack. "Is that the source? Johnson's timber yard?"

He nodded. "For sure. But look here."

I looked where he pointed. We were still in the graveyard, approaching the tumbledown end wall, beyond which stood the derelict shack. Running in a parallel series along the dry ground, from the mill and into the graveyard, deep cracks showed through the tangled brambles, briars, and grasses. One of these cracks, wider than the others, had actually split a heavy horizontal marble slab right down its length. Garth grunted. "That wasn't done last time I was here," he said.

"The sea's been at it again." I nodded. "Undermining the cliffs. Maybe we're not as safe here as you think."

He glanced at me. "Not the sea this time," he said, very definitely. "Something else entirely. See, there's been no rain for weeks. Everything's dry. And *it* gets thirsty same as we do. Give me a hand."

He stood beside the broken slab and got his fingers into

the crack. It was obvious that he intended to open up the tomb. "Garth," I cautioned him. "Isn't this a little ghoulish? Do you really intend to desecrate this grave?"

"See the date?" he said. "1847. Heck, I don't think he'd mind, whoever he is. Desecration? Why, he might even thank us for a little sweet sunlight! What are you afraid of? There can only be dust and bones down there now."

Full of guilt, I looked all about while Garth struggled with the fractured slab. It was a safe bet that there wasn't a living soul for miles around, but I checked anyway. Opening graves isn't my sort of thing. But having discovered him for a stubborn old man, I knew that if I didn't help him he'd find a way to do it by himself anyway; and so I applied myself to the task. Between the two of us we wrestled one of the two halves to the edge of its base, finally toppled it over. A choking fungus reek at once rushed out to engulf us! Or maybe the smell was of something else and I'd simply smelled what I "expected" to.

Garth pulled a sour face. *"Ugh!"* was his only comment.

The air cleared and we looked into the tomb. In there, a coffin just a little over three feet long, and the broken sarcophagus around it filled with dust, cobwebs, and a few leaves. Garth glanced at me out of the corner of his eye. "So now you think I'm wrong, eh?"

"About what?" I answered. "It's just a child's coffin."

"Just a little 'un, aye." He nodded. "And his little coffin looks intact, doesn't it? *But is it?*" Before I could reply he reached down and rapped with his horny knuckles on the wooden lid.

And despite the fact that the sun was shining down on us, and for all that the sea gulls cried and the world seemed at peace, still my hair stood on end at what happened next. For the coffin lid collapsed like a puffball and fell into dusty debris, and—God help me—*something in the box gave a grunt and puffed itself up into view!*

I'm not a coward, but there are times when my limbs have a will of their own. Once when a drunk insulted my wife, I struck him without consciously knowing I'd done it. It was that fast, the reaction that instinctive. And the same now. I didn't pause to draw breath until I'd cleared the wall and was halfway up the field to the paved path; and even then I probably wouldn't have stopped, except I tripped and fell flat and knocked all the wind out of myself.

By the time I stopped shaking and sat up, Garth was puffing and panting up the slope toward me. "It's all right," he was gasping. "It was nothing. Just the rot. It had grown in there and crammed itself so tight, so confined, that when the coffin caved in . . ."

He was right and I knew it. I *had* known it even with my flesh crawling, my legs, heart, and lungs pumping. But even so: "There were . . . *bones* in it!" I said, contrary to common sense. "A skull."

He drew close, sank down beside me gulping at the air. "The little un's bones," he panted, "caught up in the fibers. I just wanted to show you the extent of the thing. Didn't want to scare you to death!"

"I know, I know." I patted his hand. "But when it moved—"

"It was just the effect of the box collapsing," he explained, logically. "Natural expansion. Set free, it unwound like a jack-in-the-box. And the noise it made—"

"—That was the sound of its scraping against the rotten timber, amplified by the sarcophagus." I nodded. "I know all that. It shocked me, that's all. In fact, two hours in your bloody Easingham have given me enough shocks to last a lifetime!"

"But you see what I mean about the rot?" We stood up, both of us still a little shaky.

"Oh, yes, I see what you mean. I don't understand your

obsession, that's all. Why don't you just leave the damned stuff alone?"

He shrugged but made no answer, and so we made our way back toward his home. On our way the silence between us was broken only once. "There!" said Garth, looking back toward the brow of the hill. "You see him?"

I looked back, saw the dark outline of an Alsatian dog silhouetted against the rise. "Ben?" Even as I spoke the name, so the dog disappeared into the long grass beside the path.

"Ben!" Garth called, and blew his piercing whistle. But with no result. The old man worriedly shook his head. "Can't think what's come over him," he said. "Then again, I'm more his friend than his master. We've always pretty much looked after ourselves. At least I know that he hasn't run off. . . ."

Then we were back at Garth's house, but I didn't go in. His offer of another coffee couldn't tempt me. It was time I was on my way again. "If ever you're back this way—" he said as I got into the car.

I nodded, leaned out of my window. "Garth, why the hell don't you get out of here? I mean, there's nothing here for you now. Why don't you take Ben and just clear out?"

He smiled, shook his head, then shook my hand. "Where'd we go?" he asked. "And anyway, Lily-Anne's still here. Sometimes in the night, when it's hot and I have trouble sleeping, I can feel she's very close to me. Anyway, I know you mean well."

That was that. I turned the car round and drove off, acknowledged his final wave by lifting my hand briefly, so that he'd see it.

Then, driving round a gentle bend and as the old man sideslipped out of my rearview mirror, I saw Ben. He was crossing the road in front of me. I applied my brakes, let him get out of the way. It could only be Ben, I supposed:

a big Alsatian, shaggy, yellow eyed. And yet I caught only a glimpse; I was more interested in controlling the car, in being sure that he was safely out of the way.

It was only after he'd gone through the hedge and out of sight into a field that an afterimage of the dog surfaced in my mind: the way he'd seemed to limp—his belly hairs, so long as to hang down and trail on the ground, even though he wasn't slinking—a bright splash of yellow on his side, as if he'd brushed up against something freshly painted.

Perhaps understandably, peculiar images bothered me all the way back to London; yes, and for quite a long time after. . . .

Before I knew it a year had gone by, then eighteen months, and memories of those strange hours spent in Easingham were fast receding. Faded with them was that promise I had made myself to visit my parents more frequently. Then I got a letter to say my mother hadn't been feeling too well, and another right on its heels to say she was dead. She'd gone in her sleep, nice and easy. This last was from a neighbor of theirs: my father wasn't much up to writing right now, or much of anything else for that matter; the funeral would be on . . . at . . . etc, etc.

God!—how guilty I felt driving up there, and more guilty with every mile that flashed by under my car's wheels. And all I could do was choke the guilt and the tears back and drive and feel the dull, empty ache in my heart that I knew my father would be feeling in his. And of course that was when I remembered old Garth Bentham in Easingham, and my "advice" that he should get out of that place. It had been a cold sort of thing to say to him. Even cruel. But I hadn't known that then. I hadn't thought.

We laid Ma to rest and I stayed with the Old Man for a few days, but he really didn't want me around. I thought about saying: "Why don't you sell up, come and live with

us in London." We had plenty of room. But then I thought
of Garth again and kept my mouth shut. Dad would work
it out for himself in the fullness of time.

It was late on a cold Wednesday afternoon when I started
out for London again, and I kept thinking how lonely it
must be in old Easingham. I found myself wondering if
Garth ever took a belt or filled a pipe, if he could even
afford to, and . . . I'd promised him that if I was ever back
up this way I'd look him up, hadn't I? I stopped at an
off-license, bought a bottle of half-decent whisky and some
pipe and rolling baccy, and a carton of two hundred cigarettes and a few cigars. Whatever was his pleasure, I'd
probably covered it. And if he didn't smoke, well I could
always give the tobacco goods to someone who did.

My plan was to spend just an hour with Garth, then head
for the motorway and drive to London in darkness. I don't
mind driving in the dark, when the weather and visibility
are good and the driving lanes all but empty, and the night
music comes sharp and clear out of the radio to keep me
awake.

But approaching Easingham down that neglected cul-de-
sac of a road, I saw that I wasn't going to have any such
easy time of it. A storm was gathering out to sea, piling up
the thunderheads like beetling black brows all along the
twilight horizon. I could see continuous flashes of lightning
out there, and even before I reached my destination I could
hear the high seas thundering against the cliffs. When I did
get there—

Well, I held back from driving quite as far as the barrier,
because only a little way beyond it my headlights had
picked out black, empty space. Of the three houses that had
stood closest to the cliffs only one was left, and that one
slumped right on the rim. So I stopped directly opposite
Garth's place, gave a honk on my horn, then switched off
and got out of the car with my carrier-bag full of gifts.

Making my way to the house, the rush and roar of the sea was perfectly audible, transferring itself physically through the earth to my feet. Indeed the bleak, unforgiving ocean seemed to be working itself up into a real fury.

Then, in a moment, the sky darkened over and the rain came on out of nowhere, bitter cold and squally, and I found myself running up the overgrown garden path to Garth's door. Which was when I began to feel really foolish. There was no sign of life behind the grimy windows, neither a glimmer of light showing, nor a puff of smoke from the chimney. Maybe Garth had taken my advice and got out of it after all.

Calling his name over the rattle of distant thunder, I knocked on the door. After a long minute there was still no answer. But this was no good; I was getting wet and angry with myself; I tried the doorknob, and the door swung open. I stepped inside, into deep gloom, and groped on the wall near the door for a light switch. I found it, but the light wasn't working. Of course it wasn't: there was no electricity! This was a ghost town, derelict, forgotten. And the last time I was here it had been in broad daylight.

But . . . Garth had made coffee for me. On a gas-ring? It must have been.

Standing there in the small cloakroom shaking rain off myself, my eyes were growing more accustomed to the gloom. The cloakroom seemed just as I remembered it: several pieces of tall, dark furniture, pine-paneled inner walls, the old grandfather clock standing in one corner. Except that this time . . . the clock wasn't clucking. The pendulum was still, a vertical bar of brassy fire where lightning suddenly brought the room to life. Then it was dark again—if anything even darker than before—and the windows rattled as thunder came down in a rolling, receding drumbeat.

"Garth!" I called again, my voice echoing through the

old house. "It's me, Greg Lane. I said I'd drop in some time? . . ." No answer, just the *hiss* of the rain outside, the feel of my collar damp against my neck, and the thick, rising smell of . . . of what? And suddenly I remembered very clearly the details of my last visit here.

"Garth!" I tried one last time, and I stepped to the door of his living room and pushed it open. As I did so there came a lull in the beating rain. I heard the floorboards creak under my feet, but I also heard . . . a groan? My sensitivity at once rose by several degrees. Was that Garth? Was he hurt? *My God!* What had he said to me that time? "One of these days the postman will find me stretched out in here, and he'll think: 'Well, I needn't come out here anymore.' "

I had to have light. There'd be matches in the kitchen, maybe even a torch. In the absence of a mains supply, Garth would surely have to have a torch. Making my way shufflingly, very cautiously across the dark room toward the kitchen, I was conscious that the smell was more concentrated here. Was it just the smell of an old, derelict house, or was it something worse? Then, outside, lightning flashed again, and briefly the room was lit up in a white glare. Before the darkness fell once more, I saw someone slumped on the old settee where Garth had served me coffee. . . .

"Garth?" The word came out half-strangled. I hadn't wanted to say it; it had just gurgled from my tongue. For though I'd seen only a silhouette, outlined by the split-second flash, it hadn't looked like Garth at all. It had been much more like someone else I'd once seen—in a photograph. That drooping right shoulder.

My skin prickled as I stepped on shivery feet through the open door into the kitchen. I forced myself to draw breath, to think clearly. *If* I'd seen anyone or anything at all back there (it could have been old boxes piled on the settee, or a roll of carpet leaning there), then it most probably had been

Garth, which would explain that groan. It *was* him, of course it was. But in the storm, and remembering what I did of this place, my mind was playing morbid tricks with me. No, it was Garth, and he could well be in serious trouble. I got a grip of myself, quickly looked all around.

A little light came into the kitchen through a high back window. There was a two-ring gas cooker, a sink, and a drainer board with a drawer under the sink. I pulled open the drawer and felt about inside it. My nervous hand struck what was unmistakably a large box of matches, and—yes, the smooth heavy cylinder of a hand torch!

And all the time I was aware that someone was or might be slumped on a settee just a few swift paces away through the door to the living room. With my hand still inside the drawer, I pressed the stud of the torch and was rewarded when a weak beam probed out to turn my fingers pink. Well, it wasn't a powerful beam, but any sort of light had to be better than total darkness.

Armed with the torch, which felt about as good as a weapon in my hand, I forced myself to move back into the living room and directed my beam at the settee. But oh, Jesus—all that sat there was a monstrous grey mushroom! It was a great fibrous mass, growing out of and welded with mycelium strands to the settee, and in its center an obscene yellow fruiting body. But for God's sake, it had the shape and outline and *look* of an old woman, and it had Lily-Anne's deflated chest and slumped shoulder!

I don't know how I held onto the torch, how I kept from screaming out loud, why I simply didn't fall unconscious. That's the sort of shock I experienced. But I did none of these things. Instead, on nerveless legs, I backed away, backed right into an old wardrobe or Welsh dresser. At least, I backed into what had once *been* a piece of furniture. But now it was something else.

Soft as sponge, the thing collapsed and sent me sprawl-

ing. Dust and (I imagined) dark red spores rose up everywhere, and I skidded on my back in shards of crumbling
wood and matted webs of fiber. And lolling out of the
darkness behind where the dresser had stood—bloating out
like some loathsome puppet or dummy—a second fungoid
figure leaned toward me. And this time it was a caricature
of Ben!

He lolled there, held up on four fiber legs, muzzle snarling soundlessly, for all the world tensed to spring—and all
he was was a harmless fungous thing. And yet this time I
did scream. Or I think I did, but the thunder came to drown
me out.

Then I was on my feet, and my feet were through the
rotten floorboards, and I didn't care except I had to get out
of there, out of that choking, stinking, collapsing—

I stumbled, *crumbled* my way into the tiny cloakroom,
tripped and crashed into the clock where it stood in the
corner. It was like a nightmare chain reaction that I'd
started and couldn't stop; the old grandfather just crumpled
up on itself, its metal parts clanging together as the wood
disintegrated around them. And all the furniture following
suit, and the very wall paneling smoking into ruin where I
fell against it.

And there where that infected timber had been, there he
stood—old Garth himself! He leaned half out of the wall
like a great nodding manikin, his entire head a livid yellow
blotch, his arm and hand making a noise like a huge puffball bursting underfoot where they separated from his side
to point floppingly toward the open door. I needed no more
urging.

"God! Yes! *I'm going!"* I told him, as I plunged out into
the storm. . . .

After that . . . nothing, not for some time. I came to in a
hospital in Stokesley about noon the next day. Apparently

I'd run off the road on the outskirts of some village or other, and they'd dragged me out of my car where it lay upside down in a ditch. I was banged up and so couldn't do much talking, which is probably as well.

But in the newspapers I read how what was left of Easingham had gone into the sea in the night. The churchyard, Haitian timber, terrible dry-rot fungus, the whole thing, sliding down into the sea and washed away forever on the tides.

And yet now I sometimes think:

Where did all that wood *go* that Garth had been selling for years? And what of all those spores I'd breathed and touched and rolled around in? And sometimes when I think things like that it makes me feel quite ill.

I suppose I shall just have to wait and see. . . .

# The Man Who Photographed Beardsley

*This one first appeared in a Hugh Lamb anthology in 1976. That was only four years after my first book,* The Caller of the Black, *had been released by Arkham House. But in his brief introduction, Hugh was already including me among his list of "famous names." Also, I seem to remember him saying somewhere that the story was "unique." But aren't they all? Anyway, his comment pleased and surprised me at the time, and still does.*

*It strikes me that with a bit of effort I could rewrite this and make it really splatter. But I don't think I will. . . .*

Gentlemen, my very own darling boys in blue:
If as you seem so determined to avow, there has been about my work an element of the grotesque and macabre ('criminal' I simply *refuse* to admit) and if, in order to achieve the perfection of ultimate realism, I have indeed allowed myself to become a "fanatic" and my art an "obsession" . . . Why! Is it any wonder? I find in these things, in your assertions, nothing to excite any amazement or stupefaction. Nothing, that is, other than the natural astonishment and fascination of my subject matter.

And yet, in his own day, Beardsley's work excited just

such—yes, horror! "His own day!" An inadequate cliché, that, for of course he had no "day"; his work is as fresh and inspiring now—and possibly more so, as witness the constant contemporary cannibalization of his style and plagiarism of his techniques—as when first he put ink to paper. His art exists in a time-defying limbo of virtuosity that will parallel, I am sure, the very deathlessness of Tut-ankhamen as glimpsed in the gold of his funerary mask.

Times change, my dears! Does Aubrey Vincent Beardsley's art excite any such powerfully outraged emotions now? Except in the most naïve circles, it does *not!* Admiration, yes—delight and fascination, of course—but disgust, revulsion . . . horror? No. Nor, I put it to you, will *my* art fifty years hence. No, certainly not in the finished, perfect article; neither in that nor in the contemplation of its controversial construction. I would think it likely that there will be controversy; but in the end, though my name be vilified initially, I will be exonerated *through my art itself!* Yes, in the end my art must win.

I digress? My words are irrelevant? My apologies, my dears, it is the genius in me. Genius, like truth, will out.

I started, in what now seems a mundane manner, void of any airy aspiration or ambition other than that of exercising my art to the best of my ability, by selling pornographic photographs to the men's magazines. Now: I say "pornographic", and yet even the most permissive of my pictures—so erotic as to shock even, well, Beardsley himself—were executed with a finesse such as to abnegate any prohibitive reactions from my publishers, and they were undeniably the delight of my public.

Came the time when I could afford to be more choosey in my work, when I commanded the attentions of all the first-line publishers in the field, and then I would accept only the choicest contracts. Eventually I was sufficiently established to quit working to order completely, and then I was able to give more time and energy to my own projects. To my delight, my new photographic experiments went down well, better in fact than even the best of my previous works. Pornography became a thing of the past as I grew ever the more fascinated with the exotic, the weird, the outré.

Following the Beardsley Project I had plans for Hieronymus Bosch, but—

Pardon? I must try not to wander? The Beardsley Project? . . . Once more I apologize. Well, gentlemen, it came about like this:

I had long been disillusioned with color. Colors annoyed me. Without motion they somehow looked untrue, and they certainly did nothing to enhance the sombre quality of my more sepulchral pieces. Oh, the greens and purples were all right, when I could get just the right lighting effects— Rhine Castles and Infernal Caverns and so on—but as for the rest of the spectrum I couldn't give a damn! And yet Beardsley had ignored color without sacrificing an iota of feeling, in perfect perspective, with the result that his black-and-whites are the wonders of intricate design that they are. But if I was to take a lead from Beardsley in tone, why not in texture? Why not in my actual material?

It was then that I began to photograph Beardsley.

At first there was no pattern in my mind, no plot as to the development of my theme; I simply photographed him the way I found him. My *Venus Between Terminal Gods* became an instant success, as was *The Billet-Doux*. Indeed in the latter picture I actually beat A.V. at his own game! The absolute intricacy of my work was a wonder to behold. The delicate frills and ribbons of her negligee, the art nouveau of the headboard, the perfect youthful beauty of the girl herself . . .

Ah, but *The Barge* was my pièce de résistance. Oh, I was already well on the road toward that type of success I

craved, but *The Barge* made me. It took me six days to set
the thing up, to get the studio balcony absolutely *right,* to
have the costumes perfected and find the perfectly skeletal
gent (profile) to match the young woman (the girl from *The
Billet-Doux)* with the great fan. And the *wigs,* my dears,
you'd never believe the trouble I—

But there . . . I mustn't go on, must I?

That was when Nigel Naith of *Fancy* asked me for a
series. A complete series, prepaid and entirely in my own
time, and I literally had carte blanche. The one stipulation
Nigel did make—and it made me really *furious*—was that
there should be color. *Well!* . . .

I determined to do the entire series on Beardsley the
Weird, commencing with *Alberich!* The trouble I had finding just the right shape of ugly, shrunken dwarf! I went on
to *Don Juan, Sagnarelle, and the Beggar*—and it must be
immediately apparent that my difficulty in obtaining a decent beggar was enormous. In the end, to obtain a skull of
such loathly proportions and contours, I turned to a local
mental institution with the tale that I was preparing a photographic documentary of the unfortunate inmates of such
refuges.

Fortunately the custodians of the place did not know my
work; eventually I was allowed to borrow one Stanley, who
was allegedly quite servile, for my purpose. And he was
absolutely ideal—except that he didn't like having to dress
in rags. I had to beat him, and it was necessary to dub the
two normal models in for they simply refused to pose with
him; and moreover he *bit* me, but in the end I had my
picture.

Pardon? Well I *know* I'm taking my time, sweetheart, but
it's my story, isn't it?

Yes . . . Well, the first real difficulty came when I was
working on the last picture of the set. Yes, *that* picture, *The
Dancer's Reward.* Now the costumery and props were fairly

easy—even that wafer-thin, paint-pallet tabletop, with its single supporting hairy leg—but I knew that the central piece, the head, was going to give me problems. Ah, that head!

And so I looked around for a model. He or she—I'm not fussy, my darlings—would have to have long black hair, that much was obvious. And I would like, too, someone of a naturally pale complexion. . . . Naturally.

Eventually I found him, a young dropout from up north; long black hair, pale complexion, rather gaunt; he was the one, yes. And he was short of money, which of course was important. I managed to get hold of him on his first day in town, before he made any friends, which was also important. That first night he stayed at my place, but it worked out he wasn't my type, which was just as well in the circumstances. I mean, well, I couldn't afford to get emotional about him, now could I?

The next morning, bright and early, we were up and about. I took him straight downstairs to the studio, unlocked the door, and let him in. Then I lit the joss sticks, sat him down in a studio chair, and went out to get the morning milk from the doorstep. That was so I could make us both a cup of coffee right there in the studio. Coffee steadies my nerves, you know? . . .

When I got back in he was complaining about the smell. Well, of course, that put me on my guard. I told him I liked the studio to smell sweet, you know, and pointed out all the air fresheners I had about the place and the sprays I used. And I explained away the joss sticks by telling him that the smell of incense gave the place the right sort of atmosphere. He had me a bit worried, though, when he asked me what I kept in the back room. I mean, I just couldn't *tell* him, now could I?

Anyway, I set up a three-minute timed exposure on my studio camera and then kept one eye on my watch. While

I was waiting I took a few dummy shots of the lad against various backdrops; getting him in the mood, you know? Then I had him lean over the workbench with his arms wide apart, staring straight ahead. He was getting fifteen a shot—he thought—and so he was willing to oblige; you might even say overjoyed. I took a few more dummy shots from the front, then moved round to the side.

So there we were: me clicking away with my little camera, and no film in it and all, and him all stretched out over the workbench staring ahead.

"That's it!" I kept saying, and "Just look in front there," and "Money for old rope!" Stuff like that. And click, click, click with the empty camera. And I moved behind him and got the cleaver from behind the curtains, and his eyes had just started to swivel round when—

I got him first time, and clean as a whistle, which was just as well for I didn't want to mess the neck up. Believe me, it was quick. You've seen the groove in the bench? And still a full minute to go.

On with my wig, with all its tight little ringlets; and the costume, all pinned up just so. Then over to the table with its hairy leg and ornate band. The blood slopping; the head tilted back just so; my left hand held thus and my right holding his forelocks; the slippers in exactly the right position. And, my dears, his mouth fell open *of its own accord!* Beautiful! . . .

And that was when I realized that in my excitement I had made a dreadful mistake. Such a silly little thing really: when I brought in the milk I forgot to lock the outside door. The studio door, too, for I'd been distracted by the lad's complaints about the smell. And that, of course, was all it took to undo me.

It was the new postman: a nosey parker, just like the old one. In through the studio door he came, waving a pink envelope that could only contain a letter from Nigel Naith;

and when he *saw* what lay across the workbench!—*And what* I had in my hand—

Well, I couldn't let him get away, now could I, my dears? No, of course not.

But he did get away—he did! Oh, I managed to grab the cleaver all right, before he'd even moved. I mean he was still standing there all gasping and white, you know? But damn me if that costume of mine didn't let me down! Halfway across the room I tripped on the thing and went over like a felled oak; at which the postman seemed to come to life again, let out one *terrific* scream, ran for it.

And so there was nothing left for me to do but read Nigel's letter while I waited for you dears to come. Poor Nigel: when was I going to send him the goodies, he wanted to know? Ah, but there'll be no more of my work in *Fancy,* I fear.

What's that, my love? The back room? Yes, yes, of course you're right. The joss sticks? Yes, of *course,* my dear. And the—thing—on the bed? Ah, but now, I really must protest. That was a *model* of mine! "Thing," indeed!

Yes, yes, a model, something else I was working on. The theme? Why, *Edgar Allan Poe's Illustrators,* my darling. Harry Clarke was the artist, and—

You do? Why you clever boy! Yes, of course it was *Valdemar,* and—

Who was he? Why, the old postman, who else! The first one, yes. Of course he still had another week or so to go to reach a proper state of—

You've got all you wanted? Well, anything to help the law, I always say. But isn't it a shame about Nigel—I mean about him paying me in advance and all?

Tell me: is the *Police Gazette* a glossy?

# The Man Who Felt Pain

*"The Man Who Felt Pain" was something of a departure for me
. . . from true horror, that is. Which is to say, it's more SF
than horror. But it does have macabre implications, and not
only for "The Man." In any case, I find the borders between
horror, fantasy, and SF increasingly difficult to define, and I
think it entirely possible that readers are having the same
trouble. The readers of* Fantasy Tales, *where this first saw
the light of day, voted it the best story in that issue. Which
perhaps says something for the value of cross-pollination.*

B ut, you would ask, don't we all?
    Yes, I would answer, we all feel pain—our own, and
perhaps a little of those who are closest to us—but rarely
anyone else's. We don't physically feel *everyone* else's pain.
My twin brother Andrew felt everyone's pain, or would
have if he'd been able to bear it, but of course he couldn't
and in the end it killed him. Yes, and now it would kill me,
too, except I intend to put myself way, way beyond it.

So what do I mean, he could feel everyone's pain? Do I
mean he was a man of God, who felt *for* people? A man
who agonized over all the world's strife and turmoil, who
felt the folly and frustration of men maiming and killing

each other in their petty squabbles and wars? Well it's true he did, to a certain extent, but no, that isn't what I mean. I mean that he was the next leap forward in the evolution of the human race. I mean that he was a member of time's tiny fraternity of genuine geniuses, *sui generis* in fact, until the day he died. If he had happened on the shores of some primal ocean, then he could have been the Missing Link; or five million years ago he might have been the first ape-man to use a branch to lever rocks down in an avalanche upon his next meal; or a million years later employed fire to cook that meal; or just two million years ago used the first log "wheels" to roll a megalith boulder to and fro across the entrance to his cave. They were all steps forward, and so was Andrew, except he was a leap.

For if we *all* felt everyone's pain, why, then there'd be no more wars or cruelties or hurtfulness of any sort and we could get on with the real business of our being here— which is to question *why* we're here, and to care for each other, and to go on . . . wherever.

I've thought about it a lot up here, where there's plenty of space and time to think, and my thoughts have been diverse.

There are these green bushes (I forget their name) that have oval leaves in tight, mathematically precise rows down their stems, and if you hold a burning match under one of them they *all* close up! And not only on that bush but on every other bush of that species in the vicinity! An intricate trigger mechanism created by Nature—or God if you're a believer—and transmitted through sap and fiber, branch, twig, root, and perhaps even soil; intricate and yet simple, if you know how. A card up the sleeve of . . . of a bush?

In the ocean there are polyps—organisms, occasionally huge, made up of tiny single individual units each with lives of their own—that, when the predator fish bites one, the *entire colony* retracts into the safety of its alveolate rock or anchorage. Nature has allowed each to feel the agony of the others—for self-preservation. But to give such a gift to . . . a coral? A jellyfish? A polyp? If it could be done for such lowly creatures as these, why then create Man and simply leave him to his own devices? Surely that were to ask for trouble!

And so Andrew was the next step forward, for when he was born Nature also gave the gift to him. Except that I saw it in action and know that in fact it was a curse.

Now from up here I look down on the world revolving far below—at the beautiful green and blue planet Earth, which is slowly but surely destroying me—and while I remember almost exactly how it began, I daren't even think how it will end. . . .

Our mother was American, our father English, and we were born in August 2027 at Lyon, France, where at that time could be found the headquarters of ESP, the European Space Programme. Our parents worked on the Programme: she was and still is a computer technician, and he a PTI and instructor astronaut. He had journeyed into space many times during that decade in which we were born, but was forced to give it up when the technology got beyond him. A pity he never had Mother's mental wizardry, her computer-oriented brain. Anyway he has a desk job now, from which he'll retire, but reluctantly, in another five or six years' time.

I suppose it was only natural that Andrew and I should want to be astronauts: by the time Dad was finishing up we were already cramming math and computer studies, aviation and astronautics, spaceflight subjects across the board. And like the twins we were—like peas in a pod—we paralleled each other in performance. If I was top of the class one term, Andrew would pip me the next, and vice versa. At nineteen we flew the ESP shuttles (pilot and/or copilot, whichever task suited us at the time, or simply as crew members) and at twenty-one we'd been to Moonbase and back. Always together.

The trouble started at Cannes, South of France, in the summer of 2049, when we were resting up after a month-long series of shuttle runs to destroy a lot of outdated space debris: sputs and sats and bits of old rockets lodged in their many, often dangerous orbits up there far outside Earth's envelope. I won't go into details, for any ten-year-old kid knows them: it was just a matter of giving these odd piles of freewheeling, obsolete junk a little shove in the right direction at the right time, to send them tumbling sadly and yet somehow grandly out and away and down into the hot heart of Sol.

But we were very young men and space is a lonely place, and so when we had our feet on the ground we liked to look for company. Nothing permanent, for we didn't lead the sort of life that makes for lasting relationships, but if you're an astronaut and can't find a little female company on a beach in Cannes . . . then it has to be time to see a plastic surgeon! On this occasion, however, we were on our own, just lying there on our towels on the beach and absorbing the heat of that especially hot summer, when it happened. I say "it" for at first we didn't know what it was. Not for quite some little time, in fact.

"Aaaah-*ow!*" said Andrew, abruptly sitting up and rapidly blinking his eyes, staring out across an entirely placid ocean. And though there was a twinge of pain in his voice, he wasn't holding himself; he'd simply gone a little pale and shuddery, as if he had stomach cramp or something.

"Ow?" I repeated him, but not quite, because the sound he'd made hadn't really been repeatable: more an animal cry than a word proper. "You were stung?"

He frowned, looked at the sand all about, shook his head. "I . . . I don't think so," he finally, uncertainly said.

I looked at him—at the physical fact and presence of my brother—in admiration, which was nice because I was looking at a better than mirror image of myself! Andrew, with his mass of gleaming black hair, blue eyes, and clean, strong features, and his athlete's body. How many times had I wondered: *Do I really look as good as this?*

But . . . a few minutes later and his stab of unknown pain was forgotten, and a spearfisherman came out of the sea with a silver-glistening fish, shot through the head, stone dead on his spear. He took off his swim fins and marched proudly off up the beach with his catch. And Andrew's eyes following him, still frowning. That was all there was to it, that first time.

After that the pains came thick and fast: big hurts and small ones, pains that made him burn or ache or sometimes simply cramped him, but occasionally agonies that doubled him over and caused him to throw up on the floor. None of them coming for any good reason that we could think of, and not a one from any visible cause or having any viable cure.

The Programme medics all agreed that there was nothing wrong with Andrew, at least not with his body, and they were the best in the world and should know. But he and I, we knew that there was something desperately wrong with him. He was feeling pain, and feeling it when in fact he was in the peak of condition and nothing, absolutely *nothing,* should hurt.

I remember a fight in a nightclub in Paris; though we weren't involved personally, still I had to carry Andrew to our car and drive him to a friend's house. It was as though he was the one who took the hammering—and not a mark on him, and anyway the scrap had taken place on the other side of the room. But he'd certainly jerked upright out of his seat, grunting and yelling and slamming this way and that as the shouting and sounds of fists striking flesh reached us!

And he'd just as surely crashed over onto his back on the floor, groggy as a punch-drunk ex-boxer, as the fight came to a close.

I remember the night in Lyon when he woke up hoarsely screaming his agony and clawing at his face. We were sleeping on the base at the time and there'd been some party or other we hadn't attended. But I'd heard the crash outside at the same time Andrew started yelling, and when I looked out of the window there was this accident down there, where a once-pretty girl had been tossed through a windshield onto the hood of the second car, her face shattered and bloody. Andrew sat on his bed and moaned and shuddered and held his face together (which *was* together, you understand) until an ambulance came and took the injured girl away. . . .

And that was when it finally began to dawn on us just what was wrong, and what was rapidly getting worse; so that it's hardly surprising he had his breakdown. He had it because he'd begun to realize that nothing and no one could ever put his problem right, and that from now on he was subject to anyone else's, and everyone else's pain.

For that was the simple fact of it: that he felt pain. From the pinprick stings of small, damaged, or dying creatures to the screaming agonies of hideous human death. But once we knew what it was, at least we could tell the doctors.

It didn't take them long to check it out, and after they did . . . I've never seen so many intelligent down-to-earth men looking so downright shocked and disbelieving and lost for answers. And *lost* is the only word for it, for how can you treat someone for the aches and pains and bumps and cuts and bruises of someone else? How can you treat— or begin treating—the agony of a broken leg when the leg plainly isn't broken?

Nonaddictive painkillers, obviously . . .

. . . Oh, really?

For in fact it did no good to give painkillers to Andrew. The pain wasn't actually in him; its source or sources were beyond his mind and body, coming from outside of him; there was nothing they could put *inside* of him that would help. Worse, it didn't even bring relief when they gave the pills to the ones actually suffering from the pain! They only *thought* the pain had gone away, because it had been blocked. But the cause of the hurting was still there and Andrew could feel it. . . .

The thing's progress was rapid; it precisely paralleled Andrew's deterioration. Obviously, he wasn't going out into space anymore. . . .

. . . Or *was* he?

Once they'd accepted this new thing—Andrew's . . . disease?—the ESP medics were amenable to an idea of mine. And they backed us on it. For seven years we'd been using one-man weather sats for accurate forecasting. The robot sats had been fine in their day, but nothing was as clear-sighted as human eyes and nothing so observant as an alert human brain. And what with the extensive damage done to the ozone layer—the constant fluctuation of its tears and holes—computer probability was at best mechanical guess-work anyway.

So . . . my idea was simple and I don't think I need to restate it. It would mean Andrew would be completely iso-lated for two months at a stretch, which isn't good for anyone, but at least it would give him time to get himself back together again before they brought him down for his periodic visits in hell. And it would also give the medics time to try to find a new angle of approach. Because if this was a disease connected with or perhaps even springing from space, then it was something they were going to have to take a crack at.

It took some haggling (the Programme Chiefs like to have 100 percent–fit men up there), but between the medics, myself, and my parents we convinced the upper echelon that Andrew should become WWO&A, a World Weather Observer & Adviser. And he and I spent another three months getting him back on his feet again, mentally and physically. Which wasn't easy.

It meant spending a lot of time in the loneliest places in the world: in deserts, on frozen ocean strands, in the wilds of Canada and blustery Scottish highlands, finally on the uninhabited beaches of Cyprus, which the deteriorating ozone layer had put paid to as far back as 2006. There weren't a hell of a lot of Venuses on half shells floating ashore at Paphos this time around.

We talked and trained, and Andrew got himself together and faced up to it, and away from all the pains of men he gradually improved and became fit again. But at the same time he'd been growing ever more aware of a very worrying thing: the PE was wearing down. PE was our jargon for the ratio between a person in pain and his distance from Andrew, the receiver. The Proximity Effect. Previously, the source had needed to be pretty close. But now . . . all the world's pain, however muted, was getting there, was getting through to him. He felt it like you might hear the sea in a shell: as a distant tumult. A roaring that was gradually creeping up on him.

Nor was that the whole thing; for he'd also become more sensitive to the agonies of the smaller creatures, whose myriad ravages were grown that much more sharp to him. A huge cloud of desiccated, exhausted migratory butterflies spiraled down out of the aching Mediterranean sky to drown in the tideless sea, and Andrew gaped and gasped and began to turn blue before the last of them had expired. He felt the dull shuddering of the tiny clam devoured by the starfish, and the intolerable burning of the stranded man-o'-war evaporating on the sand. And now he couldn't get back into space fast enough.

Except . . . he never made it.

It was on every Vidscreen in the world and dominated every newscast for a month: the blowup at Fatu Hiva in the Marquesas.

There were two launches scheduled for that day. The first was a French relief team going up to Luna Orbital Station, and the second was supposed to be Andrew shuttling up to W-Sat III. But the French team never got off the pad, which meant that Andrew never got *on* it. We were only a mile away from that mess, waiting out the countdown when it fireballed—and my twin brother felt every poor sod of them frying! If they'd all gone up at once in the bang it would have been bad enough—but three of them, blazing, managed to eject. And Andrew blazed with them.

The medics took him back then and called in the shrinks, too, and I found myself excluded. Now it had to be up to the specialists, because I couldn't reach him anymore. He'd gone "inside" and wasn't coming out for a while.

We were twins and I loved him; I might easily have gone to pieces myself, if the Old Folks had let me. But they didn't. "You've earned a lot of money, son, you and Andrew," my father told me. "Which is just as well because your brother is going to need it. Oh, I know, there are a lot of good people working on him for free—but there are other specialists who haven't even seen him yet, and they cost money. Money doesn't last forever, Ray—it comes and it goes. If you want to do something for Andrew, want to take care of his future, then the best thing would be to get yourself back out into space. Let me and your mother look after this end for a while."

Andrew's future! It hadn't even got through to the Old Folks that he didn't have one. It was something they couldn't allow themselves to believe, and so they didn't. But at least their advice was good and kept me together. I went back into space, and up there where I could look down and see everything clearly (so clearly that I used to believe it allowed me to think more clearly, too) I'd sometimes wonder: *Why him and not me? We're twins, so how come it skipped me?* But even in space there was no answer to that. Not then . . .

I did two months on W-Sat III standing in for Andrew, and almost without pause a further three months on the vast, incredible wheel that was Luna Orbital, watching the EV engineers laboriously putting together the miracle that would one day become Titan Station. And finally it was back to Earth.

Meanwhile, I hadn't been out of touch: I got coded radio mail that my personal receiver unscrambled onto discs for me. The Old Folks kept me in the picture regards Andrew.

"We found a specializing chemist who designed a drug for him," my mother told me, her languid American drawl still very much in evidence for all that she'd been expatriate for thirty years. "It has side effects—makes his whole skin itch and upsets his balance a little—but it does cut down on the pain. And it's nonaddictive!" Fine for anyone else; but my brother, my double, the athlete who was my twin? In private I cried about it.

"He's out of dock," the Old Man's gravelly English tones cheerfully informed me toward the end of one message, "house hunting off Land's End!"

That last had me stumped. What the hell was "off" Land's End? I called up the atlas on my computer and got the answer: the Isles of Scilly. But it was the wrong answer. There were also several lighthouses.

When I got back down I had three months accumulated R and R and plenty to do with it, but first the Programme Officer I/C wanted to see me. In Lyon I went up to Jean-Pierre Durant's office and was ushered in. Durant was a short, sturdy man in his fifties, wide as a door, short-

cropped greying hair, big hard hands, very powerful looking. And he *was* powerful in every way; but bighearted with
it, a man who loved his fellow men. Right then, however, I
had a down on ESP because of Andrew (to me, they'd
seemed too eager to write him off) and possibly it showed
in my face. Also, I was in a hurry to get across the Channel
to England, and down to Land's End, and out to see my
brother in the old deserted lighthouse he'd made his home.
So Durant was the Big Boss—so what? I considered this an
intrusion into my time. And perhaps that showed, too.

"Sit down, Ray," Durant invited, smiling, waving me
into a chair. He spoke English, which his accent made warm
and compassionate, salving a little of the anger and frustration out of me. "And don't worry," he continued, "I don't
intend to waste your time. I'll get right down to it: we think
there's maybe something we can do for Andrew—if it's at
all possible."

My heart gave a leap and I started to my feet again. "The
medics have come up with something?"

Durant shook his head, pointed at the chair. Frowning,
I sat. "The psychoanalysts!" I burst out again, leaning forward. "It was psychosomatic, right? Some kind of mental
allergy?"

"Ray," he said, again shaking his head, "they're working
on those things—and getting nowhere fast. And Andrew
isn't helping by making it hard for anyone to see him.
So . . . we're not making much progress. Not along those
lines, anyway."

I was still frowning. "So how can you help him?"

Durant looked tugged two ways; he sighed, shrugged,
stroked his chin. "Personally, I think he should go back out
into space again."

I stared at him for a moment, then slumped. "We tried
that," I said, disappointed.

He ignored my expression and my answer, and said:

"Way out in space." But it was *how* he said it. This time there was no way I could remain seated; I jumped up, leaned forward across his desk. If Durant meant what I thought he meant . . . it had always been our wildest dream!

"Titan?" I finally got it out.

He nodded, and: *"Way* out!" repeated himself. "Far beyond the influence of whatever it is that's killing him. If we can get him up to Moonbase for a year . . . we think we may have the Titan hardware ready by then. You've been up on the Luna Orbiter and know how hard they're all working up there. The Titan wheel was going to be unmanned at first, as you know, with its life-supports on green just waiting for a crew when we were ready to send them. However—" And he smiled again and shrugged.

I took a pace back, collapsed into my chair, dazedly shook my head as a mixture of emotions flooded through me. "But . . . why tell me? I mean, haven't you told Andrew?"

Now the smile, a worried one at best, left his face. "I told him last week—by letter, special delivery, a jet-copter—and his answer . . . wasn't satisfactory. I told him yesterday, and when he could bring himself to answer the phone I got the same response. And I've tried to tell him again this morning, but apparently he's not taking calls. So maybe you'd better tell him for me."

"Unsatisfactory?" Over everything else he'd said that one word had stuck in my mind. "His answer was unsatisfactory? In what way?"

"Ray," Durant looked straight into my eyes, "your brother is convinced he's going to die—of other people's pain. He says he's given it plenty of thought and knows there's no way of stopping it. And he says that since it's coming, he'd prefer it came here on Earth than out there. Going out into space would only delay it anyway, he says. So you're his last chance. Possibly he's already too far gone physically for the job, in which case you'll not only have to talk him into accepting it, but also get him back up on his feet one last time. You did it before, between you, so maybe you can do it again. That's the whole thing, and that's why I sent for you. . . ."

"Do my parents—?" I started, but he cut me off.

"Your parents are your parents, Ray. I know them almost as well as you do. In some respects I know them better. Andrew has forbidden them to visit him, says not to baby him and that he's doing fine, and when he's ready to see them he'll turn up on their doorstep. Do you think it's likely—or even right—that I should tell them he's going downhill? But I have told them we *might* send him out to Saturn if he wants it, and that it's up to him. Though in fact it now looks like it's up to you. When are you seeing him?"

"Tomorrow," I said. "As soon as I can get there. Right now, if there was any quick way."

"There is," he told me. "Get your things together, whatever you want to take with you, and be at the helipad in one hour. I'll clear it and see that you're jet-coptered over. Two and a half hours and you're there, OK?"

And of course I said yes, that was OK. . . .

I didn't try to call Andrew first; it was to be a surprise, and it was. But on the way across I talked to my pilot, the one who'd taken Durant's letter to Andrew on Perring's Rock. "How did he look to you?" I asked him.

Josh Bertin was a Belgian and had been a jet-copter pilot for ESP as long as I'd been around; I knew him personally and he knew our history. "Andrew . . . wasn't his brightest," he answered, carefully. And before I could quiz him further: "You know why he bought the Rock, of course?"

"Oh, yes." I nodded. "Miles out to sea. No people. No pain. Not so much, anyway."

Josh glanced at me out of the corner of his eye. "Yes . . .

and no," he said. "Oh, that's the reason he settled there, for sure, but—"

When his pause threatened to go on indefinitely, I prompted him: "But?"

"He mentioned something you and he call the PE? Something to do with how close people were to him? Well, he told me it's breaking down. All the way down."

"Josh," I was really alarmed now, "I think you'd better tell me—"

*"But,"* he broke in on me, "he's coping with it—so far. Learning to live with it. All he has to do is keep telling himself it's not real, that's all—that the pain belongs to someone else—and then he'll be OK. As long as nothing big happens. But right out there in the sea? Well, he's not expecting any disasters, you know? And Ray, that's it. No good asking me any more, 'cos that's all he told me."

I said nothing but simply turned over what he'd said in my mind. And while I was still turning it over, that's when the pain hit me. Andrew's pain—and I knew it!

It came from outside of me, slamming into me like an explosive shell and fragmenting deep inside. It was like a tankful of pain had overflowed into my guts. Someone was crushing my heart, yanking it this way and that, trying to tear it out of me. I had thought I knew what pain was, but I hadn't. *This* was pain! Big Pain!

It would have driven me surging to my feet, but I was strapped in. I cried out, or gurgled, and then I must have blacked out. . . .

When I came to Josh had slapped an oxygen mask over my nose and mouth and was shaking me. He'd switched the jet-copter to automatic pilot, and he was white as death. But as soon as I opened my eyes, dragged the mask off my face, and let it fall, then he took a deep breath and climbed

down a little. "Are you OK?" he said. And: "Jesus, Ray—
what was all *that?*"

At that time I'd known what it was, but now I couldn't
be sure, didn't want to be sure. I had thought it was Andrew, something from him that couldn't be contained, overflowing into me. But . . . I didn't even know if that was
possible. Being a twin I knew all about the so-called "Corsi-
can Brothers" cases, but nothing like that had ever happened to me (to us?) before. So . . . maybe it was just me.
My heart? Had I been pushing it too hard?

"I don't know what it was," I finally answered Josh. "I'm
too scared to *think* what it was. I only know it was pain, and
that it's gone now."

But I didn't tell him that something else had gone, too,
something that I hadn't even been aware until suddenly—
right there and then—I no longer had it. It had been a warm
feeling, that's all. A feeling that there was something out
there other than what I could see, feel, and touch. A sure
knowledge that the universe was bigger than me. Now that
I'd lost it I knew that it had been something greater than
merely "I think, therefore I am." Perhaps it had been *"we
think. . . ."* But now there was just an emptiness, with
nothing out there at all except the world and all of space
and all the other stars and worlds in it. And for the first time
in my life I experienced loneliness. Even with someone right
there beside me, I was lonely. . . .

It was mid-September, still warm but very soggy, and fog
lay like a milky shroud on the ocean where Perring's Rock
stuck up like a partly clenched fist from the grey-surging
water, its lighthouse index finger pointing at the leaden sky.
Perring's Rock was the sloping acre-and-a-half plateau of
some drowned mountain, rising seventy or eighty feet out
of the sea and having the lighthouse built at the highest
point of the slope. There was something of a tiny scalloped

bay and beach to the west, away from our approach path, and a flat area on our side of the lighthouse picked out with typical helipad patterns. Like 90 percent of all lighthouses the rock had been abandoned since before the turn of the century, when super high-tech Radar, Skyspy, and the W-Sats had put them out of business for good. But the way they'd built it, the sea wasn't going to claim this one for a long time still to come. And desolate? The place looked about as lonely as I now felt. Except as we landed I saw that it wasn't, or saw something that caused me to think that it wasn't.

It was Andrew himself!—leaning over the rail of the lighthouse's circular balcony or platform, waving to us through the blast from our fans as we came down. Then I was free of my straps and sliding the cabin door open, out of my seat and down under the rotors before they'd even nearly stopped turning, and running up the rock-carved steps to pause at the foot of the tower. And there was my brother up top, still leaning on the rail high overhead, his shirtsleeves flapping in a breeze sprung up off the sea. Except . . . he wasn't looking down at me at all but at something else. And he was so still, so very still there at the rail. Not so much leaning on it, I now saw, as propped up by it.

I was inside and up the steps three at a time; and no need to worry now about the state of my heart for I was galvanized, my actions electric, supercharged! Yanked aloft by a fear and a pain beyond physical pain, I *hurtled* up those steps, while from behind and below me Josh Bertin's cry followed despairingly from the well of the corkscrew:

"Ray! . . . Ray! . . ."

Up to the old lamp room I swept, and on up its iron ladder and through the open trapdoor onto the flat, circular roof. And there was Andrew clinging to the iron three-bar safety rail—or rather hanging on it. One foot had slipped through and jammed there, dangling in space, and the other leg was bent at the knee, propping him against an upright. His left arm lay loosely along the top rail, while his opposite shoulder and arm lolled stiffly across it, supporting his weight. With his shirtsleeves flapping like that and his head on one side, he looked like . . . like a sorry scarecrow fallen from its cross. And I saw that I was right and he hadn't in fact been looking at us as we came in but at something else, down on the beach—as he'd blindly stared at it ever since his final, killing pain had reached out to me and knocked me out during the flight.

The mist had curled away a little and now I, too, could see it there on the narrow shingle strand. A beached whale, with three great, deep crimson slashes across its spine where some liner's screw had broken its broad back!

The blood was still pumping, though very sluggishly now, sporadically; overhead the gulls wheeled and cried their excitement, like vultures waiting for the last spark to flicker low and expire; out at sea a cow and her calf stood off and spouted, and it seemed to me that over and above my own pain I could feel something of theirs, too.

. . . Until, like ice water down my back, there dawned the realization that *I could actually feel it,* and finally I knew that I was compensating for Andrew's loss. . . .

That was three months ago and since then . . . I've thought about it a lot up here, where there's plenty of time and space to think. And my thoughts have been diverse.

I watch the curving reef that is Japan appear at the rim of the mighty Pacific, and as it slides closer I can point a trembling finger at the very bay where in these same moments of time the dolphins are *still* being slaughtered in their thousands. I feel the outward rush of human agony as bombs explode in Zambia, while the African continent slips by so distantly beneath my observation ports that my eyes see nothing but its beauty. A million babies are born and their mothers cry out, and a million men die—but they only feel their own pain while I feel something of all of it. And with every revolution I feel more.

Nine months to go, and Saturn is waiting for me out there beyond the pain of the world. But now and then I ask myself: will the wash from the world one day reach out to me even there. Or will I have moved on, outward to the stars, before then?

Sometimes I wonder: Are there other men or beings out there, in the stars?

. . . And sometimes I pray there are not. . . .

# The Viaduct

*The older I get, the more afraid I become of heights. More
specifically, of climbing and cliffs and being too close to the
edge. What's the difference? Well, fifteen years ago when
Ramsey Campbell asked me to write him a story for* Super-
horror, *I was very much into hang gliding, which I did at
every opportunity all over Scotland's Pentland Hills. To this
day I enjoy the rather more orthodox flying of vacation trips
to the Mediterranean or the U.S.A. I feel safe knowing that
there's something more than just my fingertips holding me up:
like the outer skin and engines of an aircraft, or even the
billowing half cones of a Rogallo wing. But climbing? Why,
I get dizzy just talking to people like M. John Harrison, who
for relaxation defies gravity to shinny up rearing great over-
hangs of rock! To my mind, just thinking of falling from a
such a height is sufficiently horrific without . . . well, without
the added complications of "The Viaduct."*

Horror can come in many different shapes, sizes, and
colors; often, like death, which is sometimes its companion, unexpectedly. Some years ago horror came to two
boys in the coal-mining area of England's northeast coast.

Pals since they first started school seven years earlier,

their names were John and David. John was a big lad and thought himself very brave; David was six months younger, smaller, and he wished he could be more like John.

It was a Saturday in the late spring, warm but not oppressive, and since there was no school the boys were out adventuring on the beach. They had spent most of the morning playing at being starving castaways, turning over rocks in the life-or-death search for crabs and eels—and jumping back startled, hearts racing, whenever their probing revealed too frantic a wriggling in the swirling water, or perhaps a great crab carefully sidling away, one pincer lifted in silent warning—and now they were heading home again for lunch.

But lunch was still almost two hours away, and it would take them less than an hour to get home. In that simple fact were sown the seeds of horror, in that and in one other fact—that between the beach and their respective homes there stood the viaduct. . . .

Almost as a reflex action, when the boys left the beach they headed in the direction of the viaduct. To do this they turned inland, through the trees and bushes of the narrow dene that came right down to the sand, and followed the path of the river. The river was still fairly deep, from the spring thaw and the rains of April, and as they walked, ran, and hopped they threw stones into the water, seeing who could make the biggest splash.

In no time at all, it seemed, they came to the place where the massive, ominous shadow of the viaduct fell across the dene and the river flowing through it, and there they stared up in awe at the giant arched structure of brick and concrete that bore upon its back one hundred yards of the twin tracks that formed the coastal railway. Shuddering mightily whenever a train roared overhead, the man-made bridge was a never-ending source of amazement and wonder to them. . . . And a challenge, too.

It was as they were standing on the bank of the slow-moving river, perhaps fifty feet wide at this point, that they spotted on the opposite bank the local village idiot, "Wiley Smiley." Now of course, that was not this unfortunate youth's real name; he was Miles Bellamy, victim of cruel genetic fates since the ill-omened day of his birth some nineteen years earlier. But everyone called him Wiley Smiley.

He was fishing, in a river that had supported nothing bigger than a minnow for many years, with a length of string and a bent pin. He looked up and grinned vacuously as John threw a stone into the water to attract his attention. The stone went quite close to the mark, splashing water over the unkempt youth where he stood a little way out from the far bank, balanced none too securely on slippery rocks. His vacant grin immediately slipped from his face; he became angry, gesturing awkwardly and mouthing incoherently.

"He'll come after us," said David to his brash companion, his voice just a trifle alarmed.

"No he won't, stupid," John casually answered, picking up a second, larger stone. "He can't get across, can he." It was a statement, not a question, and it was a fact. Here the river was deeper, overflowing from a large pool directly beneath the viaduct, which, in the months ahead, children and adults alike would swim in during the hot weekends of summer.

John threw his second missile, deliberately aiming it at the water as close to the enraged idiot as he could without actually hitting him, shouting: "Yah! Wiley Smiley! Trying to catch a whale, are you?"

Wiley Smiley began to shriek hysterically as the stone splashed down immediately in front of him and a fountain of water geysered over his trousers. Threatening though they now were, his angry caperings upon the rocks looked very funny to the boys (particularly since his rage was impotent), and John began to laugh loud and jeeringly. David, not a cruel boy by nature, found his friend's laughter so infectious that in a few seconds he joined in, adding his own voice to the hilarity.

Then John stooped yet again, straightening up this time with two stones, one of which he offered to his slightly younger companion. Carried completely away now, David accepted the stone and together they hurled their missiles, dancing and laughing until tears rolled down their cheeks as Wiley Smiley received a further dousing. By that time the rocks upon which their victim stood were thoroughly wet and slippery, so that suddenly he lost his balance and sat down backward into the shallow water.

Climbing clumsily, soggily to his feet, he was greeted by howls of laughter from across the river, which drove him to further excesses of rage. His was a passion that might only find outlet in direct retaliation, revenge. He took a few paces forward, until the water swirled about his knees, then stooped and plunged his arms into the river. There were stones galore beneath the water, and the face of the tormented youth was twisted with hate and fury now as he straightened up and brandished two that were large and jagged.

Where his understanding was painfully slow, Wiley Smiley's strength was prodigious. Had his first stone hit John on the head, it might easily have killed him. As it was, the boy ducked at the last moment and the missile flew harmlessly above him. David, too, had to jump to avoid being hurt by a flying rock, and no sooner had the idiot loosed both his stones than he stooped down again to grope in the water for more.

Wiley Smiley's aim was too good for the boys, and his continuing rage was making them begin to feel uncomfortable, so they beat a hasty retreat up the steeply wooded slope of the dene and made for the walkway that was fastened and ran parallel to the nearside wall of the viaduct. Soon they had climbed out of sight of the poor soul below, but they could still hear his meaningless squawking and shrieking.

A few minutes more of puffing and panting, climbing steeply through trees and saplings, brought them up above the wood and to the edge of a grassy slope. Another hundred yards and they could go over a fence and onto the viaduct. Though no word had passed between them on the subject, it was inevitable that they should end up on the viaduct, one of the most fascinating places in their entire world. . . .

The massive structure had been built when first the collieries of the Northeast opened up, long before plans were drawn up for the major coast road, and now it linked twin colliery villages that lay opposite each other across the narrow river valley it spanned. Originally constructed solely to accommodate the railway, and used to that end to this very day, with the addition of a walkway, it also provided miners who lived in one village but worked in the other with a shortcut to their respective coal mines.

While the viaduct itself was of sturdy brick, designed to withstand decade after decade of the heavy traffic that rumbled and clattered across its triple-arched back, the walkway was a comparatively fragile affair. That is not to say that it was not safe, but there were certain dangers and notices had been posted at its approaches to warn users of the presence of at least an element of risk.

Supported upon curving metal arms—iron bars about one and one-half inches in diameter, which, springing from the brick and mortar of the viaduct wall, were set perhaps twenty inches apart—the walkway itself was of wooden planks protected by a fence five feet high. There were, however, small gaps where rotten planks had been removed and never replaced, but the miners who used the viaduct were careful and knew the walkway's dangers intimately. All in all the walkway served a purpose and was reasonably safe; one might jump from it, certainly, but only a very careless person or an outright fool would fall. Still, it was no place for anyone suffering from vertigo. . . .

Now, as they climbed the fence to stand gazing up at those ribs of iron with their burden of planking and railings, the two boys felt a strange, headlong rushing emotion within them. For this day, of course, was *the* day!

It had been coming for almost a year, since the time when John had stood right where he stood now to boast: "One day I'll swing hand over hand along those rungs, all the way across. Just like Tarzan." Yes, they had sensed this day's approach, almost as they might sense Christmas or the end of long, idyllic summer holidays . . . or a visit to the dentist. Something far away, which would eventually arrive, but not yet.

Except that now it had arrived.

"One hundred and sixty rungs," John breathed, his voice a little fluttery, feeling his palms beginning to itch. "Yesterday, in the playground, we both did twenty more than that on the climbing-frame."

"The climbing-frame," answered David, with a naïve insight and vision far ahead of his age, "is only seven feet high. The viaduct is about a hundred and fifty."

John stared at his friend for a second and his eyes narrowed. Suddenly he sneered. "I might have known it—you're scared, aren't you?"

"No," David shook his head, lying, "but it'll soon be lunchtime, and—"

"You *are* scared!" John repeated. "Like a little kid. We've been practicing for months for this, every day of school on the climbing-frame, and now we're ready. You know we can do it." His tone grew more gentle, urging:

"Look, it's not as if we can't stop if we want to, is it? There's them holes in the fence, and those big gaps in the planks."

"The first gap," David answered, noticing how very far away and faint his own voice sounded, "is almost a third of the way across. . . ."

"That's right," John agreed, nodding his head eagerly. "We've counted the rungs, haven't we? Just fifty of them to that first wide gap. If we're too tired to go on when we get there, we can just climb up through the gap onto the walkway."

David, whose face had been turned toward the ground, looked up. He looked straight into his friend's eyes, not at the viaduct, in whose shade they stood. He shivered, but not because he was cold.

John stared right back at him, steadily, encouragingly, knowing that his smaller friend looked for his approval, his reassurance. And he was right, for despite the fact that their ages were very close, David held him up as some sort of hero. No daredevil, David, but he desperately wished he could be. And now . . . here was his chance.

He simply nodded—then laughed out loud as John gave a wild whoop and shook his young fists at the viaduct. "Today we'll beat you!" he yelled, then turned and clambered furiously up the last few yards of steep grassy slope to where the first rung might easily be reached with an upward spring. David followed him after a moment's pause, but not before he heard the first arch of the viaduct throw back the challenge in a faintly ringing, sardonic echo of John's cry: "Beat you . . . beat you . . . beat you . . ."

As he caught up with his ebullient friend, David finally allowed his eyes to glance upward at those skeletal ribs of iron above him. They looked solid, were solid, he knew—but the air beneath them was very thin indeed. John turned to him, his face flushed with excitement. "You first," he said.

"Me?" David blanched. "But—"

"You'll be up onto the walkway first if we get tired," John pointed out. "Besides, I go faster than you—and you wouldn't want to be left behind, would you?"

David shook his head. "No," he slowly answered, "I wouldn't want to be left behind." Then his voice took on an anxious note: "But you won't hurry me, will you?"

" 'Course not," John answered. "We'll just take it nice and easy, like we do at school."

Without another word, but with his ears ringing strangely and his breath already coming faster, David jumped up and caught hold of the first rung. He swung forward, first one hand to the rung in front, then the other, and so on. He heard John grunt as he too jumped and caught the first rung, and then he gave all his concentration to what he was doing.

Hand over hand, rung by rung, they made their way out over the abyss. Below them the ground fell sharply away, each swing of their arms adding almost two feet to their height, seeming to add tangibly to their weight. Now they were silent, except for an occasional grunt, saving both breath and strength as they worked their way along the underside of the walkway. There was only the breeze that whispered in their ears and the infrequent toot of a motor's horn on the distant road.

As the bricks of the wall moved slowly by, so the distance between rungs seemed to increase, and already David's arms felt tired. He knew that John, too, must be feeling it, for while his friend was bigger and a little stronger, he was also heavier. And sure enough, at a distance of only twenty-five, maybe thirty rungs out toward the center, John breathlessly called for a rest.

David pulled himself up and hung his arms and his rib cage over the rung he was on—just as they had practiced in the playground—getting comfortable before carefully turning his head to look back. He was shocked to see that John's face was paler than he'd ever known it, that his eyes were staring. When John saw David's doubt, however, he managed a weak grin.

"It's OK," he said. "I was—I was just a bit worried about you, that's all. Thought your arms might be getting a bit tired. Have you—have you looked down yet?"

"No," David answered, his voice mouselike. *No,* he said again, this time to himself, *and I'm not going to!* He carefully turned his head back to look ahead, where the diminishing line of rungs seemed to stretch out almost infinitely to the far side of the viaduct.

John had been worried about him. Yes, of course he had; that was why his face had looked so funny, so—shrunken. John thought he was frightened, was worried about his self-control, his ability to carry on. Well, David told himself, he had every right to worry; but all the same he felt ashamed that his weakness was so obvious. Even in a position like that, perched so perilously, David's mind was far more concerned with the other boy's opinion of him than with thoughts of possible disaster. And it never once dawned on him, not for a moment, that John might really only be worried about himself. . . .

Almost as if to confirm beyond a doubt the fact that John had little faith in his strength, his courage—as David hung there, breathing deeply, preparing himself for the next stage of the venture—his friend's voice, displaying an unmistakable quaver, came to him again from behind:

"Just another twenty rungs, that's all, then you'll be able to climb up onto the walkway."

*Yes,* David thought, *I'll be able to climb up. But then I'll know that I'll never be like you—that you'll always be better than me—because you'll carry on all the way across!* He set his teeth and dismissed the thought. It wasn't going to be like that, he told himself, not this time. After all, it was no different up here from in the playground. You were only higher, that was all. The trick was in not looking down—

As if obeying some unheard command, seemingly with a morbid curiosity of their own, David's eyes slowly began to turn downward, defying him. Their motion was only arrested when David's attention suddenly centered upon a spiderlike dot that emerged suddenly from the cover of the trees, scampering frantically up the opposite slope of the valley. He recognized the figure immediately from the faded blue shirt and black trousers that it wore. It was Wiley Smiley.

As David lowered himself carefully into the hanging position beneath his rung and swung forward, he said: "Across the valley, there—that's Wiley Smiley. I wonder why he's in such a hurry?" There had been something terribly *urgent* about the idiot's quick movements, as if some rare incentive powered them.

"I see him," said John, sounding more composed now. "Huh! He's just an old nutter. My dad says he'll do something one of these days and have to be taken away."

"Do something?" David queried, pausing briefly between swings. An uneasiness completely divorced from the perilous game they were playing rose churningly in his stomach and mind. "What kind of thing?"

"Dunno," John grunted. "But anyway, don't—*uh!*—talk."

It was good advice: don't talk, conserve wind, strength, take it easy. And yet David suddenly found himself moving faster, dangerously fast, and his fingers were none too sure as they moved from one rung to the next. More than once he was hanging by one hand while the other groped blindly for support.

It was very, very important now to close the distance between himself and the sanctuary of the gap in the planking. True, he had made up his mind just a few moments ago to carry on beyond that gap—as far as he could go before admitting defeat, submitting—but all such resolutions were gone now as quickly as they came. His one thought was of climbing up to safety.

It had something to do with Wiley Smiley and the eager, *determined* way he had been scampering up the far slope. Toward the viaduct. Something to do with that, yes, and with what John had said about Wiley Smiley being taken away one day . . . for *doing* something. David's mind dared not voice its fears too specifically, not even to itself. . . .

Now, except for the occasional grunt—that and the private pounding of blood in their ears—the two boys were silent, and only a minute or so later David saw the gap in the planking. He had been searching for it, sweeping the rough wood of the planks stretching away overhead anxiously until he saw the wide, straight crack that quickly enlarged as he swung closer. Two planks were missing here, he knew, just sufficient to allow a boy to squirm through the gap without too much trouble.

His breath coming in sobbing, glad gasps, David was just a few rungs away from safety when he felt the first tremors vibrating through the great structure of the viaduct. It was like the trembling of a palsied giant. "What's that?" he cried out loud, terrified, clinging desperately to the rung above his head.

"It's a—*uh!*—train!" John gasped, his own voice now very hoarse and plainly frightened. "We'll have to—*uh!*—wait until it's gone over."

Quickly, before the approaching train's vibrations could shake them loose, the boys hauled themselves up into positions of relative safety and comfort, perching on their rungs beneath the planks of the walkway. There they waited and shivered in the shadow of the viaduct, while the shuddering rumble of the train drew ever closer, until, in a protracted clattering of wheels on rails, the monster rushed by unseen overhead. The trembling quickly subsided and the train's distant whistle proclaimed its derision; it was finished with them.

Without a word, holding back a sob that threatened to develop into full-scale hysteria, David lowered himself once more into the full-length hanging position; behind him, breathing harshly and with just the hint of a whimper escaping from his lips, John did the same. Two, three more forward swings and the gap was directly overhead. David looked up, straight up to the clear sky.

"Hurry!" said John, his voice the tiniest whisper. "My hands are starting to feel funny. . . ."

David pulled himself up and balanced across his rung, tremulously took away one hand and grasped the edge of the wooden planking. Pushing down on the hand that grasped the rung and hauling himself up with the other, finally he kneeled on the rung and his head emerged through the gap in the planks. He looked along the walkway . . .

. . . There, not three feet away, legs widespread and eyes burning with a fanatical hatred, crouched Wiley Smiley. David saw him, saw the pointed stick he held, felt a thrill of purest horror course through him. Then, in the next instant, the idiot lunged forward and his mouth opened in a demented parody of a laugh. David saw the lightning movement of the sharpened stick and tried to avoid its thrust. He felt the point strike his forehead just above his left eye and fell back, off balance, arms flailing. Briefly his left hand made contact with the planking again, then lost it, and he fell with a shriek . . . across the rung that lay directly beneath him. It was not a long fall, but fear and panic had already winded David; he simply closed his eyes and sobbed, hanging on for dear life, motionless. But only for a handful of seconds.

Warm blood trickled from David's forehead, falling on

his hands where he gripped the rung. Something was prodding the back of his neck, jabbing viciously. The pain brought him back from the abyss, and he opened his eyes to risk one sharp, fearful glance upward. Wiley Smiley was kneeling at the edge of the gap, his stick already moving downward for another jab. Again David moved his head to avoid the thrust of the stick, and once more the point scraped his forehead.

Behind him David could hear John moaning and screaming alternately: "Oh, Mum! Dad! It's Wiley Smiley! It's him, him, *him!* He'll kill us, kill us." . . . Galvanized into action, David lowered himself for the third time into the hanging position and swung forward, away from the inflamed idiot's deadly stick. Two rungs, three, then he carefully turned about face and hauled himself up to rest. He looked at John through the blood that dripped slowly into one eye, blurring his vision.

David blinked to clear his eye of blood, then said: "John, you'll have to turn round and go back, get help. He's got me here. I can't go forward any farther, I don't think, and I can't come back. I'm stuck. But it's only fifty rungs back to the start. You can do it easy, and if you get tired you can always rest. I'll wait here until you fetch help."

"Can't, can't, *can't,*" John babbled, trembling wildly where he lay half across his rung. Tears ran down the older boy's cheeks and fell into space like salty rain. He was deathly white, eyes staring, frozen. Suddenly yellow urine flooded from the leg of his short trousers in a long burst. When he saw this, David, too, wet himself, feeling the burning of his water against his legs but not caring. He felt very tiny, very weak now, and he knew that fear and shock were combining to exhaust him.

Then, as a silhouette glimpsed briefly in a flash of lightning, David saw in his mind's eye a means of salvation. "John," he urgently called out to the other boy. "Do you remember near the middle of the viaduct? There are two gaps close together in the walkway, maybe only a dozen or so rungs apart."

Almost imperceptibly, John nodded, never once moving his frozen eyes from David's face. "Well," the younger boy continued, barely managing to keep the hysteria out of his own voice, "if we can swing to—" Suddenly David's words were cut off by a burst of insane laughter from above, followed immediately by a loud, staccato thumping on the boards as Wiley Smiley leapt crazily up and down.

"No, no, *no*—" John finally cried out in answer to David's proposal. His paralysis broken, he began to sob unashamedly. Then, shaking his head violently, he said: "I can't move—can't move!" His voice became the merest whisper. "Oh, God—Mum—Dad! I'll fall, I'll fall!"

"You won't fall, you git—*coward!*" David shouted. Then his jaw fell open in a gasp. John, a coward! But the other boy didn't even seem to have heard him. Now he was trembling as wildly as before and his eyes were squeezed tight shut.

"Listen," David said. "If you don't come . . . then I'll leave you. You wouldn't want to be left on your own, would you?" It was an echo as of something said a million years ago.

John stopped sobbing and opened his eyes. They opened very wide, unbelieving. "Leave me?"

"Listen," David said again. "The next gap is only about twenty rungs away, and the one after that is only another eight or nine more. Wiley Smiley can't get after both of us at once, can he?"

"You go," said John, his voice taking on fresh hope and his eyes blinking rapidly. "You go and maybe he'll follow you. Then I'll climb up and—and chase him off. . . ."

"You won't be able to chase him off," said David scornfully, "not just you on your own. You're not big enough."

"Then I'll . . . I'll run and fetch help."

"What if he doesn't follow after me?" David asked. "If we both go, he's bound to follow us."

"David," John said, after a moment or two. "David, I'm . . . frightened."

"You'll have to be quick across the gap," David said, ignoring John's last statement. "He's got that stick—and of course he'll be listening to us."

"I'm frightened," John whispered again.

David nodded. "OK, you stay where you are, if that's what you want—but I'm going on."

"Don't leave me, don't leave me!" John cried out, his shriek accompanied by a peal of mad and bubbling laughter from the unseen idiot above. "Don't go!"

"I have to, or we're both finished," David answered. He slid down into the hanging position and turned about-face, noting as he did so that John was making to follow him, albeit in a dangerous, panicky fashion. "Wait to see if Wiley Smiley follows me!" he called back over his shoulder.

"No. I'm coming, I'm coming!"

From far down below in the valley David heard a horrified shout, then another. They had been spotted. Wiley Smiley heard the shouting, too, and his distraction was sufficient to allow John to pass by beneath him unhindered. From above, the two boys now heard the idiot's worried mutterings and gruntings, and the hesitant sound of his feet as he slowly kept pace with them along the walkway. He could see them through the narrow cracks between the planks, but the cracks weren't wide enough for him to use his stick.

David's arms and hands were terribly numb and aching by the time he reached the second gap, but seeing the gloating, twisted features of Wiley Smiley leering down at him he ducked his head and swung on to where he was once more protected by the planks above him. John had stopped short of the second gap, hauling himself up into the safer, resting position.

Above them Wiley Smiley was mewling viciously like a wild animal, howling as if in torment. He rushed crazily back and forth from gap to gap, jabbing uselessly at the empty air between the vacant rungs. The boys could see the bloodied point of the stick striking down first through one open space, then the other. David achingly waited until he saw the stick appear at the gap in front of him and then, when it retreated and he heard Wiley Smiley's footsteps hurrying overhead, swung swiftly across to the other side. There he turned about to face John, and with what felt like his last ounce of strength pulled himself up to rest.

Now, for the first time, David dared to look down. Below, running up the riverbank and waving frantically, were the antlike figures of three men. They must have been out for a Saturday morning stroll when they'd spotted the two boys hanging beneath the viaduct's walkway. One of them stopped running and put his hands up to his mouth. His shout floated up to the boys on the clear air: "Hang on, lads, hang on!"

"Help!" David and John cried out together, as loud as they could. "Help!—Help!"

"We're coming, lads," came the answering shout. The men hurriedly began to climb the wooded slope on their side of the river and disappeared into the trees.

"They'll be here soon," David said, wondering if it would be soon enough. His whole body ached and he felt desperately weak and sick.

"Hear that, Wiley Smiley?" John cried hysterically, staring up at the boards above him. "They'll be here soon—and then you'll be taken away and locked up!" There was no answer. A slight wind had come up off the sea and was carrying a salty tang to them where they lay across their rungs.

"They'll take you away and lock you up," John cried again, the ghost of a sob in his voice; but once more the only answer was the slight moaning of the wind. John looked across at David, maybe twenty-five feet away, and said: "I think . . . I think he's gone." Then he gave a wild shout. "He's gone. *He's gone!*"

"I didn't hear him go," said David, dubiously.

John was very much more his old self now. "Oh, he's gone, all right. He saw those men coming and cleared off. David, I'm going up!"

"You'd better wait," David cried out as his friend slid down to hang at arm's length from his rung. John ignored the advice; he swung forward hand over hand until he was under the far gap in the planking. With a grunt of exertion, he forced the tired muscles of his arms to pull his tired body up. He got his rib cage over the rung, flung a hand up, and took hold of the naked plank to one side of the gap, then—

In that same instant David sensed rather than heard the furtive movement overhead. "John!" he yelled. "He's still there—*Wiley Smiley's still there!*"

But John had already seen Wiley Smiley; the idiot had made his presence all too plain, and already his victim was screaming. The boy fell back fully into David's view, the hand he had thrown up to grip the edge of the plank returning automatically to the rung, his arms taking the full weight of his falling body, somehow sustaining him. There was a long gash in his cheek from which blood freely flowed.

"Move forward!" David yelled, terror pulling his lips back in a snarling mask. "Forward, where he can't get at you. . . ."

John heard him and must have seen in some dim, frightened recess of his mind the common sense of David's advice. Panting hoarsely—partly in dreadful fear, partly from hideous emotional exhaustion—he swung one hand forward and caught at the next rung. And at that precise moment, in the split second while John hung suspended between the two rungs with his face turned partly upward, Wiley Smiley struck again.

David was witness to it all. He heard the maniac's rising, gibbering shriek of triumph as the sharp point of the stick lanced unerringly down, and John's answering cry of purest agony as his left eye flopped bloodily out onto his cheek, lying there on a white thread of nerve and gristle. He saw John clap *both hands* to his monstrously altered face, and watched in starkest horror as his friend seemed to stand for a moment, defying gravity, on the thin air. Then John was gone, dwindling away down a drafty funnel of air, while rising came the piping, diminishing scream that would haunt David until his dying day, a scream that was cut short after what seemed an impossibly long time.

John had fallen. At first David couldn't accept it, but then it began to sink in. His friend had fallen. He moaned and shut his eyes tightly, lying half across and clinging to his rung so fiercely that he could no longer really feel his bloodless fingers at all. John had fallen. . . .

Then—perhaps it was only a minute or so later, perhaps an hour, David didn't know—there broke in on his perceptions the sound of clumping, hurrying feet on the boards above, and a renewed, even more frenzied attack of gibbering and shrieking from Wiley Smiley. David forced his eyes open as the footsteps came to a halt directly overhead. He heard a gruff voice:

"Jim, you keep that bloody—*Thing*—away, will you? He's already killed one boy today. Frank, give us a hand here."

A face, inverted, appeared through the hole in the planks not three feet away from David's own face. The mouth opened and the same voice, but no longer gruff, said: "It's okay now, son. Everything's okay. Can you move?"

In answer, David could only shake his head negatively. Overtaxed muscles, violated nerves had finally given in. He was frozen on his perch; he would stay where he was now until he was either taken off physically or until he fainted.

Dimly the boy heard the voice again, and others raised in an urgent hubbub, but he was too far gone to make out any words that were said. He was barely aware that the face had been withdrawn. A few seconds later there came a banging and tearing from immediately above him; a small shower of tiny pieces of wood, dust, and homogenous debris fell upon his head and shoulders. Then daylight flooded down to illuminate more brightly the shaded area beneath the walkway. Another board was torn away, and another.

The inverted face again appeared, this time at the freshly made opening, and an exploratory hand reached down. Using its kindly voice, the face said: "OK, son, we'll have you out of there in a jiffy. I—*uh!*—can't quite seem to reach you, but it's only a matter of a few inches. Do you think you can—"

The voice was cut off by a further outburst of incoherent shrieking and jabbering from Wiley Smiley. The face and hand withdrew momentarily and David heard the voice yet again. This time it was angry. "Look, see if you can keep that damned idiot back, will you? And keep him quiet, for God's sake!"

The hand came back, large and strong, reaching down. David still clung with all his remaining strength to the rung, and though he knew what was expected of him—what he must do to win himself the prize of continued life—all sense of feeling had quite gone from his limbs and even shifting his position was a very doubtful business.

"Boy," said the voice, as the hand crept inches closer and the inverted face stared into his, "if you could just reach up your hand, I—"

"I'll—I'll try to do it," David whispered.

"Good, good," his would-be rescuer calmly, quietly answered. "That's it, lad, just a few inches. Keep your balance, now."

David's hand crept up from the rung, and his head, neck, and shoulder slowly turned to allow it free passage. Up it tremblingly went, reaching to meet the hand stretching down from above. The boy and the man, each peered into the other's straining face, and an instant later their fingertips touched—

There came a mad shriek, a frantic pounding of feet, and cries of horror and wild consternation from above. The inverted face went white in a moment and disappeared, apparently dragged backward. The hand disappeared, too. And that was the very moment that David had chosen to free himself of the rung and give himself into the protection of his rescuer. . . .

He flailed his arms in a vain attempt to regain his balance. Numb, cramped, cold with that singular icy chill experienced only at death's positive approach, his limbs would not obey. He rolled forward over the bar and his legs were no longer strong enough to hold him. He didn't even feel the toes of his shoes as they struck the rung—the last of him to have contact with the viaduct—before his fall began. And if the boy thought anything at all during that fall, well, those thoughts will never be known. Later he could not remember.

Oh, there was to be a later, but David could hardly have believed it while he was falling. And yet he was not unconscious. There were vague impressions: of the sky, the looming arch of the viaduct flying past, trees below, the sea on the horizon, then the sky again, all slowly turning. There was a composite whistling, of air displaced and air ejected from lungs contracted in a high-pitched scream. And then, it seemed a long time later, there was the impact. . . .

But David did not strike the ground . . . he struck the

pool. The deep swimming hole. The blessed, merciful river!
He had curled into a ball—the fetal position, almost—
and this doubtless saved him. His tightly curled body entered the water with very little injury, however much of a
splash it caused. Deep as the water was, nevertheless David
struck the bottom with force, the pain and shock awakening whatever facilities remained functional in the motor
areas of his brain. Aided by his resultant struggling, however weak, the ballooning air in his clothes bore him surely
to the surface. The river carried him a few yards down-
stream to where the banks formed a bottleneck for the pool.

Through all the pain David felt his knees scrape pebbles,
felt his hands on the mud of the bank, and where willpower
presumably was lacking, instinct took over. Somehow he
crawled from the pool, and somehow he hung on grimly to
consciousness. Away from the water, still he kept on crawling, as from the horror of his experience. Unseeing, he
moved toward the towering unconquered colossus of the
viaduct. He was quite blind as of yet; there was only a red,
impenetrable haze before his bloodied eyes; he heard nothing but a sick roaring in his head. Finally his shoulder
struck the bole of a tree that stood in the shelter of the
looming brick giant, and there he stopped crawling,
propped against the tree.

Slowly, very slowly the roaring went out of his ears, the
red haze before his eyes was replaced by lightning flashes
and kaleidoscopic shapes and colors. Normal sound suddenly returned with a great pain in his ears. A rush of wind
rustled the leaves of the trees, snatching away and then
giving back a distant shouting that seemed to have its
source overhead. Encased in his shell of pain, David did not
immediately relate the shouting to his miraculous escape.
Sight returned a few moments later, and he began to cry
wrackingly with relief; he had thought himself permanently
blind. And perhaps even now he had not been completely

wrong, for his eyes had plainly been knocked out of order. Something was—*must be*—desperately wrong with them.

David tried to shake his head to clear it, but this action only brought fresh, blinding pain. When the nausea subsided he blinked his eyes, clearing them of blood and peering bewilderedly about at his surroundings. It was as he had suspected: the colors were all wrong. No, he blinked again, some of them seemed perfectly normal.

For instance: the bark of the tree against which he leaned was brown enough, and its dangling leaves were a fresh green. The sky above was blue, reflected in the river, and the bricks of the viaduct were a dull orange. Why then was the grass beneath him a lush red streaked with yellow and grey? Why was this unnatural grass wet and sticky, and—

—*And why were these tatters of dimly familiar clothing flung about in exploded, scarlet disorder?*

When his reeling brain at last delivered the answer, David opened his mouth to scream. Fainting before he could do so, he fell face down into the sticky embrace of his late friend.

# Recognition

*And now we're back to bumps in the night, and to Lovecraft.*
*This one owes its background to HPL's Cthulhu Mythos, and*
*its existence to W. Paul Ganley, who was looking for stories*
*for* Weirdbook *at the time. When I wrote it I was mindful of*
*Paul's penchant for the old sting in the tale. Or should that*
*be tail? Whichever, this story has a twist in the proboscis.*

## I

"As to why I asked you all to join me here, and why
I'm making it worth your while by paying each of
you five hundred pounds for your time and trouble, the
answer is simple. The place appears to be haunted, and I
want rid of the ghost."

The speaker was young, his voice cultured, his features
fine and aristocratic. He was Lord David Marriot, and the
place of which he spoke was a Marriot property: a large,
ungainly, mongrel architecture of dim and doubtful origins,
standing gaunt and gloomily atmospheric in an acre of
brooding oaks. The wood itself stood central in nine acres
of otherwise barren moors borderland.

Lord Marriot's audience numbered four: the sprightly

octogenarian Lawrence Danford, a retired man of the cloth; by contrast the so-called "mediums" Jonathan Turnbull and Jason Lavery, each a "specialist" in his own right; and myself, an old friend of the family whose name does not really matter since I had no special part to play. I was simply there as an observer—an adviser, if you like—in a matter for which, from the beginning, I had no great liking.

Waiting on the arrival of the others, I had been with David Marriot at the old house all afternoon. I had long known something of the history of the place . . . and a little of its legend. There I now sat, comfortable and warm as our host addressed the other three, with an excellent sherry in my hand while logs crackled away in the massive fireplace. And yet suddenly, as he spoke, I felt chill and uneasy.

"You two gentlemen," David smiled at the mediums, "will employ your special talents to discover and define the malignancy, if indeed such an element exists; and you, sir," he spoke to the elderly cleric, "will attempt to exorcise the unhappy—creature?—once we know who or what it is." Attracted by my involuntary agitation, frowning, he paused and turned to me. "Is something troubling you, my friend? . . ."

"I'm sorry to have to stop you almost before you've started, David," I apologized, "but I've given it some thought and—well, this plan of yours worries me."

Lord Marriot's guests looked at me in some surprise, seeming to notice me for the first time, although of course we had been introduced; for after all they were the experts while I was merely an observer. Nevertheless, and while I was never endowed with any special psychic talent that I know of (and while certainly, if ever I had been, I never would have dabbled), I did know a little of my subject and had always been interested in such things.

And who knows?—perhaps I do have some sort of sixth sense, for as I have said I was suddenly and quite inexplicably chill with a sensation of foreboding that I knew had nothing at all to do with the temperature of the library. The others, for all their much-vaunted special talents, apparently felt nothing.

"My plan worries you?" Lord Marriot finally repeated. "You didn't mention this before."

"I didn't know before just how you meant to go about it. Oh, I agree that the house requires some sort of exorcism, that something is quite definitely wrong with the place, but I'm not at all sure that you should concern yourself with finding out *exactly* what it is you're exorcising."

"Hmm, yes, I think I might agree." Old Danford nodded his grey head. "Surely the essence of the, *harumph,* matter, is to be rid of the thing—whatever it is. Er, not," he hastily added, "that I would want to do these two gentlemen out of a job—however much I disagree with, *harumph,* spiritualism and its trappings." He turned to Turnbull and Lavery.

"Not at all, sir," Lavery assured him, smiling thinly. "We've been paid in advance, as you yourself have been paid, regardless of results. We will therefore—*perform*—as Lord Marriot sees fit. We are not, however, spiritualists. But in any case, should our services no longer be required—" He shrugged.

"No, no question of that," the owner of the house spoke up at once. "The advice of my good friend here has been greatly valued by my family for many years, in all manner of problems, but he would be the first to admit that he's no expert in matters such as these. I, however, am even less of an authority, and my time is extremely short; I never have enough time for anything! That is why I commissioned him to find out all he could about the history of the house, in order to be able to offer you gentlemen something of an insight into its background.

"And I assure you that it's not just idle curiosity that

prompts me to seek out the source of the trouble here. I wish to dispose of the property, and prospective buyers just will not stay in it long enough to appreciate its many good features! And so if we are to lay something to rest here, something which ought perhaps to have been laid to rest long ago, then I want to know what it is. Damn me, the thing's caused me enough trouble!

"So let's please have no more talk about likes and dislikes or what should or should not be done. It will be the way I've planned it." He turned again to me. "Now if you'll be so good as to simply outline the results of your research?"

"Very well." I shrugged in acquiescence. "As long as I've made my feelings in the matter plain . . ." Knowing David the way I did, further argument would be quite fruitless; his mind was made up. I riffled through the notes lying in my lap, took a long pull on my pipe, and commenced:

"Oddly enough, the house as it now stands is comparatively modern, no more than two hundred and fifty years old; but it was built upon the shell of a far older structure, one whose origin is extremely difficult to trace. There are local legends, however, and there have always been chroniclers of tales of strange old houses. The original house is given brief mention in texts dating back almost to Roman times, but the actual site had known habitation—possibly a druidic order or some such—much earlier. Later it became part of some sort of fortification, perhaps a small castle, and the remnants of earthworks in the shape of mounds, banks, and ditches can be found even today in the surrounding countryside.

"Of course the present house, while large enough by modern standards, is small in comparison with the original; it's a mere wing of the old structure. An extensive cellar—a veritable maze of tunnels, rooms, and passages—was discovered during renovation some eighty years ago, when first the Marriots acquired the property, and then several clues were disclosed as to its earlier use.

"This wing would seem to have been a place of worship, of sorts, for there was a crude altar stone, a pair of ugly, fontlike basins, a number of particularly repugnant carvings of gargoyles or 'gods,' and other extremely ancient tools and bric-a-brac. Most of this incunabula was given into the care of the then curator of the antiquities section of the British Museum, but the carved figures were defaced and destroyed. The records do not say why. . . .

"But let's go back to the reign of James the First.

"Then the place was the seat of a family of supposed nobility, though the line must have suffered a serious decline during the early years of the seventeenth century—or perhaps fallen foul of the authorities or the monarch himself—for its name simply cannot be discovered. It would seem that for some reason, most probably serious dishonor, the family name has been erased from all contemporary records and documents!

"Prior to the fire that razed the main building to the ground in 1618, there had been a certain intercourse and intrigue of a similarly undiscovered nature between the nameless inhabitants, the de la Poers of Exham Priory near Anchester, and an obscure esoteric sect of monks dwelling in and around the semiruined Falstone Castle in Northumberland. Of the latter sect: they were wiped out utterly by northern raiders—a clan believed to have been outraged by the 'heathen activities' of the monks—and the ruins of the castle were pulled to pieces, stone by stone. Indeed, it was so well destroyed that today only a handful of historians could ever show you where it stood!

"As for the de la Poers: well, whole cycles of ill-omened myth and legend revolve around that family, just as they do about their Anchester seat. Suffice it to say that in 1923 the priory was blown up and the cliffs beneath it dynamited, until the deepest roots of its foundations were obliterated. Thus the priory is no more, and the last of that line of the family is safely locked away in a refuge for the hopelessly insane.

"It can be seen then that the nameless family that lived here had the worst possible connections, at least by the standards of those days, and it is not at all improbable that they brought about their own decline and disappearance through just such traffic with degenerate or ill-advised cultists and demonologists as I have mentioned.

"Now then, add to all of this somewhat tenuously connected information the local rumors, which have circulated on and off in the villages of this area for some three hundred years—those mainly unspecified fears and old wives' tales that have sufficed since time immemorial to keep children and adults alike away from this property, off the land and out of the woods—and you begin to understand something of the aura of the place. Perhaps you can feel that aura even now? I certainly can, and I'm by no means psychic. . . ."

"Just what is it that the locals fear?" Turnbull asked. "Can't you enlighten us at all?"

"Oh, strange shapes have been seen on the paths and roads; luminous nets have appeared strung between the trees like great webs, only to vanish in daylight; and, yes, in connection with the latter, perhaps I had better mention the bas-reliefs in the cellar."

"Bas-reliefs?" queried Lavery.

"Yes, on the walls. It was writing of sorts, but in a language no one could understand—glyphs almost."

"My great-grandfather had just bought the house," David Marriot explained. "He was an extremely well-read man, knowledgeable in all sorts of peculiar subjects. When the cellar was opened and he saw the glyphs, he said they had to do with the worship of some strange deity from an obscure and almost unrecognized myth cycle. Afterward he had the greater area of the cellar cemented in—said it made the house damp and the foundations unsafe."

"Worship of some strange diety?" Old Danford spoke up. "What sort of deity? Some lustful thing that the Romans brought with them, d'you think?"

"No, older than that," I answered for Lord Marriot. "Much older. A spider-thing."

"A spider?" This was Lavery again, and he snorted the words out almost in contempt.

"Not quite the thing to sneer at," I answered. "Three years ago an aging but still active gentleman rented the house for a period of some six weeks. An anthropologist and the author of several books, he wanted the place for its solitude; and if he took to it he was going to buy it. In the fifth week he was taken away raving mad!"

"Eh? *Harumph!* Mad, you say?" Old Danford repeated me.

I nodded. "Yes, quite insane. He lived for barely six months, all the while raving about a creature named Atlach-Nacha—a spider-god from the Cthulhu Cycle of myth—whose ghostly avatar, he claimed, still inhabited the house and its grounds."

At this Turnbull spoke up. "Now really!" he spluttered. "I honestly fear that we're rapidly going from the sublime to the ridiculous!"

"Gentlemen, please!" There was exasperation now in Lord Marriot's voice. "What does it matter? You know as much now as there is to know of the history of the troubles here—more than enough to do what you've been paid to do. Now then, Lawrence—" He turned to Danford. "Have you any objections?"

"*Harumph!* Well, if there's a demon here—that is, something other than a creature of the Lord—then of course I'll do my best to help you. *Harumph!* Certainly."

"And you, Lavery?"

"Objections? No, a bargain is a bargain. I have your money, and you shall have your noises."

Lord Marriot nodded, understanding Lavery's meaning. For the medium's talent was a supposed or alleged ability to speak in the tongue of the ghost, the possessing spirit. In the event of a nonhuman ghost, however, then his mouthings might well be other than speech as we understand the spoken word. They might simply be—noises.

"And that leaves you, Turnbull."

"Do not concern yourself, Lord Marriot," Turnbull answered, flicking imagined dust from his sleeves. "I, too, would be loath to break an honorable agreement. I have promised to do an automatic sketch of the intruder, an art in which I'm well practiced, and if all goes well I shall do just that. Frankly, I see nothing at all to be afraid of. Indeed, I would appreciate some sort of explanation from our friend here—who seems to me simply to be doing his best to frighten us off." He inclined his head inquiringly in my direction.

I held up my hands and shook my head. "Gentlemen, my only desire is to make you aware of this feeling of mine of . . . yes, premonition! The very air seems to me imbued with an aura of—" I frowned. "Perhaps disaster would be too strong a word."

"Disaster?" Old Danford, as was his wont, repeated me. "How do you mean?"

"I honestly don't know. It's a feeling, that's all, and it hinges upon this desire of Lord Marriot's to know his foe, to identify the nature of the evil here. Yes, upon that, and upon the complicity of the rest of you."

"But—" the young lord began, anger starting to make itself apparent in his voice.

"At least hear me out," I protested. "Then—" I paused and shrugged. "Then . . . you must do as you see fit."

"It can do no harm to listen to him," Old Danford

pleaded my case. "I for one find all of this extremely interesting. I would like to hear his argument." The others nodded slowly, one by one, in somewhat uncertain agreement.
   "Very well," Lord Marriot sighed heavily. "Just what is
it that bothers you so much, my friend?"
   "Recognition," I answered at once. "To recognize our—
opponent?—that's where the danger lies. And yet here's
Lavery, all willing and eager to speak in the thing's voice,
which can only add to our knowledge of it; and Turnbull,
happy to fall into a trance at the drop of a hat and sketch
the thing, so that we may all know exactly what it looks
like. And what comes after that? Don't you see? The more
we learn of it, the more it learns of us!
   "Right now, this *thing*—ghost, demon, 'god,' apparition,
whatever you want to call it—lies in some deathless limbo,
extradimensional, manifesting itself rarely, incompletely in
our world. But to *know* the thing, as our lunatic anthropologist came to know it and as the superstitious villagers of these parts think they know it—that is to draw it from
its own nighted place into this sphere of existence. That is
to give it substance, to participate in its materialization!"
   "Hah!" Turnbull snorted. "And you talk of superstitious
villagers! Let's have one thing straight before we go any
further. Lavery and I do *not* believe in the supernatural, not
as the misinformed majority understand it. We believe that
there are other planes of existence, yes, and that they are
inhabited; and further that occasionally we may glimpse
alien areas and realms beyond the ones we were born to. In
this we are surely nothing less than scientists, men who have
been given rare talents, and each experiment we take part in
leads us a little farther along the paths of discovery. No
ghosts or demons, sir, but scientific phenomena that may
one day open up into whole new vistas of knowledge. Let
me repeat once more: there is nothing to fear in this, nothing at all!"

"There I cannot agree," I answered. "You must be aware, as I am, that there are well-documented cases of . . ."

"Self-hypnotism!" Lavery broke in. "In almost every case where medium experimenters have come to harm, it can be proved that they were the victims of self-hypnosis."

"And that's not all," Turnbull added. "You'll find that they were all believers in the so-called supernatural. We, on the other hand, are not. . . ."

"But what of these well-documented cases you mentioned?" Old Danford spoke up. "What sort of cases?"

"Cases of sudden, violent death!" I answered. "The case of the medium who slept in a room once occupied by a murderer, a strangler, and who was found the next morning strangled—though the room was windowless and locked from the inside! The case of the exorcist," (I paused briefly to glance at Danford), "who attempted to seek out and put to rest a certain grey thing that haunted a Scottish graveyard. Whatever it was, this monster was legended to crush its victims' heads. Well, his curiosity did for him: he was found with his head squashed flat and his brains all burst from his ears!"

"And you think that all of—" Danford began.

"I don't know what to think," I interrupted him, "but certainly the facts seem to speak for themselves. These men I've mentioned, and many others like them, all tried to understand or search for things that they should have left utterly alone. Then, too late, each of them recognized . . . something . . . and it recognized them! What *I* think really does not matter; what matters is that these men are no more. And yet here, tonight, you would commence just such an *experiment,* to seek out something you really aren't meant to know. Well, good luck to you. I for one want no part of it. I'll leave before you begin."

At that Lord Marriot, solicitous now, came over and laid

a hand on my arm. "Now you promised me you'd see this thing through with me."

"I did not accept your money, David," I reminded him.

"I respect you all the more for that," he answered. "You were willing to be here simply as a friend. As for this change of heart . . . At least stay a while and see the thing underway."

I sighed and reluctantly nodded. Our friendship was a bond sealed long ago, in childhood. "As you wish—but if and when I've had enough then you must not try to prevent my leaving."

"My word on it," he immediately replied, briskly pumping my hand. "Now then: a bite to eat and a drink, I think, another log on the fire, and then we can begin. . . ."

# II

The late autumn evening was settling in fast by the time we gathered around a heavy, circular oak table set centrally upon the library's parquet flooring, in preparation for Lavery's demonstration of his esoteric talent. The other three guests were fairly cheery, perhaps a little excited—doubtless as a result of David's plying them unstintingly with his excellent sherry—and our host himself seemed in very good spirits; but I had been little affected, and the small amount of wine I had taken had, if anything, only seemed to heighten the almost tangible atmosphere of dread that pressed in upon me from all sides. Only that promise wrested from me by my friend kept me there; and by it alone I felt bound to participate, at least initially, in what was to come.

Finally Lavery declared himself ready to begin and asked us all to remain silent throughout. The lights had been turned low at the medium's request, and the sputtering logs in the great hearth threw red and orange shadows about the spacious room.

The experiment would entail none of the usual paraphernalia beloved of mystics and spiritualists; we did not sit with the tips of our little fingers touching, forming an unbroken circle; Lavery had not asked us to concentrate or to focus our minds upon anything at all. The antique clock on the wall ticked off the seconds monotonously as the medium closed his eyes and lay back his head in his high-backed chair. We all watched him closely.

Gradually his breathing deepened and the rise and fall of his chest became regular. Then, almost before we knew it and coming as something of a shock, his hands tightened on the leather arms of his chair and his mouth began a silent series of spastic jerks and twitches. My blood, already cold, seemed to freeze at the sight of this, and I had half risen to my feet before his face grew still. Then Lavery's lips drew back from his teeth and he opened and closed his mouth several times in rapid succession, as if gnashing his teeth through a blind, idiot grin. This only lasted for a second or two, however, and soon his face once more relaxed. Suddenly conscious that I still crouched over the table, I forced myself to sit down.

As we continued to watch him, a deathly pallor came over the medium's features, and his knuckles whitened where he gripped the arms of his chair. At this point I could have sworn that the temperature of the room dropped sharply, abruptly. The others did not seem to note the fact, being far too fascinated with the motion of Lavery's exposed Adam's apple to be aware of anything else. That fleshy knob moved slowly up and down the full length of his throat, while the column of his windpipe thickened and contracted in a sort of slow muscular spasm. And at last Lavery spoke. He spoke—and at the sound I could almost feel the blood congealing in my veins!

For this was in no way the voice of a man that crackled, hissed, and gibbered from Lavery's mouth in a—language?—that surely never originated on this world or within our sphere of existence. No, it was the voice of . . . something else. Something monstrous!

Interspaced with the insane cough, whistle, and stutter of harshly alien syllables and cackling cachinnations, occasionally there would break through a recognizable combination of sounds that roughly approximated our pronunciation of "Atlach-Nacha"; but this fact had no sooner made itself plain to me than, with a wild shriek, Lavery hurled himself backward—or was *thrown* backward—so violently that he overturned his chair, rolling free of it to thrash about upon the floor.

Since I was directly opposite Lavery at the table, I was the last to attend him. Lord Marriot and Turnbull on the other hand were at his side at once, pinning him to the floor and steadying him. As I shakily joined them I saw that Old Danford had backed away into the farthest corner of the room, holding up his hands before him as if to ward off the very blackest of evils.

With an anxious inquiry I hurried toward him. He shook me off and made straight for the door.

"Danford!" I cried. "What on earth is—" But then I saw the way his eyes bulged and how terribly he trembled in every limb. The man was frightened for his life, and the sight of him in this condition made me forget my own terror in a moment.

"Danford," I repeated in a quieter tone of voice. "Are you well?"

By this time Lavery was sitting up on the floor and staring uncertainly about. Lord Marriot joined me as Danford opened the library door to stand for a moment facing us. All the blood seemed to have drained from his face; his hands fluttered like trapped birds as he stumbled backward out of the room and into the passage leading to the main door of the house.

"Abomination!" he finally croaked, and no sign of his customary "harumph!" "A presence—monstrous—ultimate abomination—*God help us! . . .*"

"Presence?" Lord Marriot repeated, taking his arm. "What is it, Danford? What's wrong, man?"

The old man tugged himself free. He seemed now somewhat recovered, but still his face was ashen and his trembling unabated. "A presence, yes," he hoarsely answered, "a monstrous presence! I could not even try to exorcise . . . *that!*" And he turned and staggered along the corridor to the outer door.

"But where are you going, Danford?" Marriot called after him.

"Away," came the answer from the door. "Away from here. I'll—I'll be in touch, Marriot—but I cannot stay here now." The door slammed behind him as he stumbled into darkness and a moment or two later came the roar of his car's engine.

When the sound had faded into distance, Lord Marriot turned to me with a look of astonishment on his face. He asked: "Well, what was that all about? Did he see something, d'you think?"

"No, David." I shook my head. "I don't think he *saw* anything. But I believe he sensed something—something perhaps apparent to him through his religious training—and he got out before it could sense him!"

We stayed the night in the house, but while bedrooms were available we all chose to remain in the library, nodding fitfully in our easy chairs around the great fireplace. I for one was very glad of the company, though I kept this fact to myself, and I could not help but wonder if the others might not now be similarly apprehensive.

Twice I awoke with a start in the huge quiet room, on both occasions feeding the red-glowing fire. And since that blaze lasted all through the night, I could only assume that at least one of the others was equally restless. . . .

In the morning after a frugal breakfast (Lord Marriot kept no retainers in the place; none would stay there, and so we had to make do for ourselves), while the others prowled about and stretched their legs or tidied themselves up, I saw and took stock of the situation. David, concerned about the aged clergyman, rang him at home and was told by Danford's housekeeper that her master had not stayed at home overnight. He had come home in a tearing rush at about nine o'clock, packed a case, told her that he was off "up north" for a few days' rest, and had left at once for the railway station. She also said that she had not liked his color.

The old man's greatcoat still lay across the arm of a chair in the library where he had left it in his frantic hurry of the night before. I took it and hung it up for him, wondering if he would ever return to the house to claim it.

Lavery was baggy eyed and disheveled and he complained of a splitting headache. He blamed his condition on an overdose of his host's sherry, but I knew for a certainty that he had been well enough before his dramatic demonstration of the previous evening. Of that demonstration, the medium said he could remember nothing; and yet he seemed distinctly uneasy and kept casting about the room and starting at the slightest unexpected movement, so that I believed his nerves had suffered a severe jolt.

It struck me that he, surely, must have been my assistant through the night; that he had spent some of the dark hours tending the fire in the great hearth. In any case, shortly after lunch and before the shadows of afternoon began to creep, he made his excuses and took his departure. I had somehow known that he would. And so three of us remained . . . three of the original five.

But if Danford's unexplained departure of the previous evening had disheartened Lord Marriot, and while Lavery's rather premature desertion had also struck a discordant note, at least Turnbull stood straight and strong on the side of our host. Despite Old Danford's absence, Turnbull would still go ahead with his part in the plan; an exorcist could always be found at some later date, if such were truly necessary. And certainly Lavery's presence was not prerequisite to Turnbull's forthcoming performance. Indeed, he wanted no one at all in attendance, desiring to be left entirely alone in the house. This was the only way he could possibly work, he assured us, and he had no fear at all about being on his own in the old place. After all, what was there to fear? This was only another experiment, wasn't it?

Looking back now I feel a little guilty that I did not argue the point further with Turnbull—about his staying alone in the old house overnight to sketch his automatic portrait of the unwanted tenant—but the man was so damned arrogant to my way of thinking, so sure of his theories and principles, that I offered no slightest opposition. So we all three spent the evening reading and smoking before the log fire, and as night drew on Lord Marriot and I prepared to take our leave.

Then, too, as darkness fell over the oaks crowding dense and still beyond the gardens, I once again felt that unnatural oppressiveness creeping in upon me, that weight of unseen energies hovering in the suddenly sullen air.

Perhaps, for the first time, Lord Marriot felt it, too, for he did not seem at all indisposed to leaving the house; indeed, there was an uncharacteristic quickness about him, and as we drove away in his car in the direction of the local village inn, I noticed that he involuntarily shuddered once or twice. I made no mention of it; the night was chill, after all. . . .

At the Traveller's Rest, where business was only moderate, we inspected our rooms before making ourselves comfortable in the snug. There we played cards until about ten o'clock, but our minds were not on the game. Shortly after ten-thirty, Marriot called Turnbull to ask if all was going well. He returned from the telephone grumbling that Turnbull was totally ungrateful. He had not thanked Lord Marriot for his concern at all. The man demanded absolute isolation, no contact with the outside world whatever, and he complained that it would now take him well over an hour to go into his trance. After that he might begin to sketch almost immediately, or he might not start until well into the night, or there again the experiment could prove to be completely fruitless. It was all a matter of circumstance, and chances would not be improved by useless interruptions.

We had left him seated in his shirtsleeves before a roaring fire. Close at hand were a bottle of wine, a plate of cold beef sandwiches, a sketch pad, and pencils. These lay upon an occasional table, which he would pull into a position directly in front of himself before sleeping, or, as he would have it, before "going into trance." There he sat, alone in that ominous old house.

Before retiring we made a light meal of chicken sandwiches, though neither one of us had any appreciable appetite. I may not speak for Marriot, but as for me . . . it took me until well into the wee small hours to get to sleep. . . .

In the morning my titled friend was at my door while I was still halfway through washing. His outward appearance was ostensibly bright and breezy, but I sensed that his eagerness to get back to the old house and Turnbull was more than simply a desire to know the outcome of the latter's experiment; he was more interested in the man's welfare than anything else. Like my own, his misgivings with regard to his plan to learn something of the mysterious and alien entity at the house had grown through the night; now he would be more than satisfied simply to discover the medium well and unharmed.

And yet what could there possibly be at the place to harm him? Again, that question.

The night had brought a heavy frost, the first of the season, and hedgerows and verges were white as from a fall of snow. Halfway through the woods, on the long gravel drive winding in toward the house, there the horror struck! Maneuvering a slight bend, Lord Marriot cursed, applied his brakes, and brought the car skidding to a jarring halt. A shape, white and grey—and hideously red—lay huddled in the middle of the drive.

It was Turnbull, frozen, lying in a crystallized pool of his own blood, limbs contorted in the agony of death, his eyes glazed orbs that stared in blind and eternal horror at a sight Lord Marriot and I could hardly imagine. A thousand circular holes of about one-half inch in diameter penetrated deep into his body, his face, all of his limbs; as if he had been the victim of some maniac with a brace and bit! Identical holes formed a track along the frosted grass verge from the house to this spot, as did Turnbull's flying footprints.

Against all my protests—weakened by nausea, white and trembling with shock as he was—still Lord Marriot raced his car the remainder of the way to the house. There we dismounted, and he entered through the door that hung mutely ajar. I would not go in with him but stood dumbly wringing my hands, numb with horror, before the leering entrance.

A minute or so later he came staggering to the door. In his hand he carried a leaf from Turnbull's sketch pad.

Before I could think to avert my eyes, he thrust the almost completed sketch toward me, crying, "Look! Look!"

I caught a glimpse of something bulbous and black, hairy and red eyed—a tarantula, a bat, a dragon—whose joined legs were tipped with sharp, chitinous darts. A mere glimpse, without any real or lasting impression of detail, and yet—

"No!" I cried, throwing up my hands before my face, turning and rushing wildly back down the long drive. "No, you fool, don't let me see it! I don't want to know! *I don't want to know!*"

# No Way Home

*"No Way Home" was written in 1974 after I finished a tour in Germany and returned to the United Kingdom (Great Britain, to all you American cousins) by car. I motored through West Germany, crossed by ferry from Bremerhaven to Harwich, then almost got lost when I came off the M1 to answer a call of nature and top up again—the car with gas and myself with coffee. OK, so it was night, I'd been out of the country a good many years, and the district was strange to me. But that's hardly an excuse. I mean, it's infuriating! How in* hell *can you get lost in your own backyard? Actually it's easy, even in your own backyard. Easier still in hell.*

If you motor up the M1 past Lanchester from London and come off before Bankhead heading west across the country, within a very few miles you enter an area of gently rolling green hills, winding country roads, and Olde Worlde villages with quaint wooden-beamed street-corner pubs and noonday cats atop leaning ivied walls. The roads there are narrow, climbing gently up and ribboning down the green hills, rolling between fields, meandering casually through woods and over brick and wooden-bridged streams; the whole background forms a pattern of peace and tranquility rarely disturbed over the centuries.

By night, though, the place takes on a different aspect. An almost miasmal aura of timelessness, of antiquity, hangs over the brooding woods and dark hamlets. The moon silvers winding hedgerows and ancient thatched roofs, and when the pubs close and the last lights blink out in farm and cottage windows, then it is as if Night had thrown her blackest cloak over the land, when even the most powerful headlight's beam penetrates the resultant darkness only with difficulty. Enough to allow you to drive, if you drive slowly and carefully.

Strangers motoring through this region—even in daylight hours—are known occasionally to lose their way, to drive the same labyrinthine lanes for hours on end in meaningless circles. The contours of the countryside often seem to defy even the most accurate sense of direction, and the roads and tracks never quite seem to tally with printed maps of the area. There are rumors almost as old as the area itself that persons have been known to pass into oblivion here—like grey smoke from cottage chimney stacks disappearing into air—never to come out again.

Not that George Benson was a stranger. True, he had not been home to England for many years—since running off as a youth, later to marry and settle in Germany—but as a boy he had known this place like the back of his hand and must have cycled for thousands of miles along dusty summer roads, lanes, and tracks, even bridle paths through the heart of this very region.

And that was why he was so perturbed now, not because of a pack of lies and fairy stories and old wives' tales heard as a boy, but because this *was* his home territory, where he'd been born and reared. Indeed, he felt more than perturbed, stupid almost. A fellow drives all the way north through Germany from Dortmund, catches the car ferry from Bremerhaven into Harwich, rolls on up-country, having made the transition from right- to left-of-the-road driving with only a very small effort . . . and then hopelessly loses himself within only a fistful of miles from home!

Anger at his own supposed stupidity turned to bitter memories of his wife and then to an even greater anger. And a hurt . . .

It didn't hurt half so much now, though, not now it was all over. But the anger was still there. And the memories of the milk of marriage gone sour. Greta had just up and left home one day. George, employing the services of a detective agency, had traced his wife to Hamburg, where he'd found her in the bed of a nightclub crooner, an old boyfriend who finally had made it good.

"Damn all Krauts!" George cursed now as he checked the speed of his car to read out the legend on a village name board. His headlights picked the letters out starkly in the surrounding darkness. "Middle Hamborough?—Never bloody heard of it!" Again he cursed as, making a quick decision, he spun the steering wheel to turn his big car about on the narrow road. He would have to start backtracking, something he hated doing because it seemed so inefficient, so wasteful. "And blast and damn all Kraut cars!" he added as his front wheels bounced jarringly onto and back off the high stone roadside curb.

"Greta!" he quietly growled to himself as he drove back down the road away from the outskirts of Middle Hamborough. "What a *bitch!*" For of course she had blamed him for their troubles, saying that she couldn't stand his meanness. Him, George Benson, mean! She simply hadn't appreciated money. She'd thought that Deutschemarks grew on trees, that pfennigs gathered like dew on the grass in the night. George, on the other hand, had inherited much of the pecuniary instincts of his father, a Yorkshireman of the Old School—and of Scottish stock to boot—who really understood the value of "brass." His old man had used to say: "Thee tak care o' the pennies, Georgie, an' the pounds'll tak care o' theysels!"

George's already pinched face tightened skull-like as his thoughts again returned to Greta. She had wanted children. Children! Damned lucky thing he had known better than to accept *that!* For God's sake, who could afford children? Then she'd complained about the food—like she'd been complaining for years—said she was getting thin because the money he gave her was never enough. But George liked his women willowy and fragile; that way there was never much fight in them. Well, he'd certainly misjudged Greta; there had been plenty of fight left in her. And their very last fight had been about food, too. He had wanted her to buy food in bulk at the supermarkets for cheapness; in turn she'd demanded a deep freezer so that the food she bought wouldn't go bad; finally George had gone off the deep end when she told him how much the freezer she had in mind would cost!

She left him that same day; moreover, she ate the last of the *wurstchen* before she went! George grinned mirthlessly as he gripped the steering wheel tighter, wishing it were Greta's scrawny neck. By God! She'd be sorry when she was fat!

Still, George had had the last laugh. Their home had been paid for fifty-fifty, but it had been in George's name. He had sold it. Likewise the furniture and the few clothes she left behind. The car had been half hers, too—but again in George's name, for Greta couldn't drive. It was all his now, his money, his car, everything. As he'd done so often in the last twenty-four hours, he took one hand from the wheel to reassuringly pat the fat wallet where its outline bulged out the upper right front of his jacket.

It was the thought of money that sent George's mind casting back an hour or so to a chance encounter at Harvey's All-Night-Grill, just off the M1. This drunk had been there—oh, a real joker and melancholy with it, too—but he had been *sooo* well-heeled! George remembered the man's queer offer: "Just show me the way home, that's all—and all I've got you can have!" And he had carried a bankbook showing a credit of over two thousand pounds. . . .

That last was hearsay, though, passed on to George by Harvey himself, the stubble-jawed, greasy-aproned owner of the place. Now that earlier accidental meeting and conversation suddenly jumped up crystal clear in George's mind. It had started when George mentioned to Harvey that he was heading for Bellington; that was when the other fellow had started to take an interest in him and had made his weird offer about being shown the way home.

God damn! George sat bolt upright behind the steering wheel. Come to think of it, he *had* heard of Middle Hamborough before! Surely that was the name of the place the drunk had been looking for—for fifteen years!

George hadn't paid much attention to the man at the time, had barely listened to his gabbled, drunken pleading. He'd passed the man off quite simply as some nut who'd heard those fanciful old rumors about people getting lost in the surrounding countryside, a drunk who was making a big play of his own personal little fantasy. The fellow would be all right when he sobered up. . . .

Now that George thought about it, though—well, why should anyone make up a story like that? And come to think of it, the man hadn't seemed all that drunk. More tired and, well, *lost,* really. . . .

Just then, cresting a low hill, as his headlights flashed across the next shallow valley, George saw the house with the big garden and the long drive winding up to it. The place stood to the right of the road, atop the next hill, and the gravel drive rose up from an ornamental stone arch and iron gate at the roadside. Dipping down the road and climbing the low hill, George read the wrought-iron legend on the gate: *HIGH HOUSE.* And now he remembered more of the—drunk's?—story.

The man had called himself Kent, and fifteen years ago, on his tenth wedding anniversary, he'd left home one morning to drive to London, there to make certain business arrangements with city-dwelling colleagues. He had taken a fairly large sum of money with him when he drove from High House, the home he himself had designed and built, which had worked out just as well for him. Turning right off the Middle Hamborough road through Meadington and onto the London road at Bankhead, Kent had driven to the city. And in London—

Kent was a partner in a building concern . . . or at least he had been. For in London he discovered that his firm had never existed, that his colleagues, Milton and Jones, while they themselves were real enough, swore they had never heard of him. "Milton, Jones & Kent" did not exist; the firm was known simply as "Milton & Jones." Not only did they not know him, they tried to have him jailed for attempted fraud!

That was only the start of it, for the real horror came when he tried to get back home—only to discover that there just wasn't any way home! George remembered now Kent's apparently drunken phrase: "A strange dislocation of space and time, a crossing of probability tracks, a passage between parallel dimensions—and a subsequent *snapping back* of space-time elastic. . . ." Only a drunk would say something like that. A drunk or a nut.

Except Harvey had insisted that Kent was sober. He was just tired, Harvey said, confused, half-mad trying to solve a fifteen-year-old problem that wasn't. . . . There had never been a Middle Hamborough, Harvey insisted. The place wasn't shown on any map; you couldn't find it in the telephone directory; no trains, buses, or roads went there. Middle Hamborough wasn't!

But Middle Hamborough *was,* George had seen it, or—

Could it be that greasy old Harvey had somehow been fooling that clown all these years, milking his money drop by drop, cashing in on some mental block or other? Or had they both simply been pulling George's leg? If so, well, it certainly seemed a queer sort of joke. . . .

George glanced at his watch. Just on 11:00 P.M.

Damn it, he'd planned to be in Bellington by now, at home with the Old Folks, and he would have been if he'd come off the M1 at the right place. Of course, when he'd left England there had been no motorway as such, just another road stretching away north and south. That was where he'd gone wrong, obviously; he'd come off the M1 too soon. He should have gone on to the next exit. Well, all right, he'd kill two birds with one stone. He'd go back to Harvey's all-nighter, check out the weird one's story again, then see if he couldn't perhaps latch on to some of the joker's change to cover his time. Then he'd try to pick up a map of the area before heading home. He couldn't go wrong with a map, now could he?

Having decided his course, and considering the winding roads and pitch-darkness, George put his foot down and sped back to Harvey's place. Parking his car, he walked through the open door into the unhealthy atmosphere and lighting of the so-called cafeteria (where the lights were kept low, George suspected to make the young cockroaches on the walls less conspicuous). He went straight to the service counter and carefully rested his elbows upon it, avoiding the splashes of sticky coffee and spilled grease. Of the equally greasy proprietor he casually inquired regarding Mr. Kent's whereabouts.

"Eh? Kent? He'll be in his room. I let him lodge here, y'know. He doesn't like to be too far from this area. . . ."

"You *let* him lodge here?" George asked, raising his eyebrows questioningly.

"Well, y'know, he pays a bit."

George nodded, silently repeating the other's words: *Yeah, I'll* bet *he pays a bit!*

"G'night there!" Harvey waved a stained dishcloth at a departing truck driver and his mate. "See y'next time." He turned back to George with a scowl. "Anyway, what's it to you, about Kent? After making y'self a quick quid or two?"

"How do you mean?" George returned, assuming a hurt look. "It's just that I think I might be able to help the poor bloke out, that's all."

"Oh?" Harvey looked suspicious. "How's that, then?"

"Well, half an hour ago I was on the road to Middle Hamborough, and I passed a place set back off the road called High House. I just thought—"

" 'Ere," Harvey cut in, a surprisingly fast hand shooting out to catch George's jacket front and pull him close so that their faces almost met across the service counter. "You tryin' t' be clever, chief?"

"Well, I'll be—" George spluttered, genuinely astonished. "What the hell d'you think you're—"

" 'Cos if you are—you an' me'll fall out, we will!"

George carefully disengaged himself. "Well," he said, "I think that answers one of my questions, at least."

"Eh? What d'you mean?" Harvey asked, still looking surly. George backed off a step.

"Looks to me like you're as mad as him, attacking me like that. I mean, I might have expected you to laugh, seeing as how I fell for your funny little joke—but I'd hardly think that you'd get all physical."

"What the 'ell are you on about?" Harvey questioned, a very convincing frown creasing his forehead. "What joke?"

"Why, about Middle Hamborough, about it not being on any map and about no roads going there and Kent looking for the place for fifteen years. I'm on about a place that's not twenty minutes' fast drive from here, signposted clear as the City of London!"

Suddenly Harvey's unwashed features paled visibly. "You mean you've actually *seen* this place?" he whispered. "And you drove past . . . High House?"

"Damn right!" George answered abruptly, feeling as though things were all unreal, a very vivid but meaningless daydream.

Harvey lifted a flap in the counter and waddled through to George's side. He was a very big man, George suddenly noticed, and the color had come back to his face with a vengeance. There was a red, angry tinge to the man's sallow features now; moreover, the cafeteria was quite empty of other souls, all bar the two of them.

"Now look here—" George blurted, as Harvey began to maneuver him into a corner.

"I shouldn't 'ave mentioned 'is money, should I?" The fat man cut him off, his piggy eyes fastening upon those of his patently intended victim, making his question more a statement than a question proper.

"See," he continued, "I'd had a couple of pints earlier, or I wouldn't 'ave let it drop about 'is predicament. 'E's been right good to me, Mr. Kent 'as—'elped me set this place up proper, 'e did—and I don't cotton to the idea of some fly-boy trying to—"

Again his arm shot out, and he grabbed Benson's throat this time, trapping him in the dim corner. "So you've been down the road to Middle Hamborough, 'ave you?—And you've seen High House, eh? Well, let me tell you, I've been looking for that place close on six years, me an' poor ol' Kent, an' not so much as a peep!

"Now I knows 'e's a bit of a nut, but I *like* 'im and we gets on fine. 'E stays 'ere, cheap like, an' we do a bit of motorin' in 'is old car—lookin' for those places you say you've seen, y'know? But we never finds 'em, an' we never

will, 'cos they're not there, see? Kent being a decent little
gent, I 'umors 'im and things is OK. But I'm no crook, if
you see what I mean, though I'm not sure I can say the same
for everybody!" He peered pointedly at George, releasing
the pressure on his windpipe enough for him to croak:

"I tell you I *have* seen High House; or, at least, I saw a
place of that name and answering the description I heard
from Kent. And I *have* been on the road to—"

*"What's that about High House?"* The question was a
hoarse, quavering whisper—hesitant, and yet filled with
excited expectancy. Hearing that whisper, Harvey immediately released his grip on Benson's neck and turned to move
over quickly to the thin, grey-haired, middle-aged man who
had appeared out of a back room behind the service
counter.

"Don't get yourself all upset, Mr. Kent," Harvey protested, holding up his hands solicitously. "It's just some
bloke tryin' to pull a fast one—"

"But I heard him say—" Kent's eyes were wide, staring
past the fat proprietor straight at Benson where he stood,
still shaken, in the corner.

George found his voice again. "I said I'd seen High
House, on the road to Middle Hamborough—and I did see
it." He shook himself, straightening his tie and shrugging
his disarranged jacket back into position. "But I didn't
come back in here to get involved with a couple of nuts.
And I don't think much of your joke."

George turned away and made for the door; then, remembering his previous trouble, he turned back to face
Harvey. "Do you have a map of the area by any chance?
I've been in Germany for some years and seem to be out of
touch. I can't seem to find my way about anymore."

For a moment Kent continued to stare very hard at the
speaker; then he half turned to Harvey. "He—he got *lost!*

And he says he's seen High House! . . . I've got to believe him; I daren't miss the chance that—"

Almost sure by now that he was the victim of some cockeyed leg-pull (and yet still experiencing niggling little subconscious doubts), George Benson shrugged. "OK. No map," he grumbled. "Well, good night, boys. Maybe I'll drop in again sometime—like next visiting day!"

"No, wait!" the thin man cried. "Do you think that you can find . . . that you can find High House again?" His voice went back to a whisper on the last half-dozen words.

"Sure, I can find it again," George told him, nodding his head. "But it's well out of my way."

"I'll make it worth your trouble," Kent quickly answered, his voice rising rapidly in what sounded to George like a bad case of barely suppressed hysteria. "I'll make it very worthwhile indeed!"

George was not the man to pass up a good thing. "My car's outside," he said. "Do you want to ride with me, or will you follow in your own car?"

"I'll ride with you. My hands are shaking so badly that I—"

"I'm coming with you," Harvey suddenly grunted, taking off his greasy apron.

"No, no, my friend." Kent turned to him. "If we don't find High House, I'll be back. Until then, and just in case we do find it, this is for all you've done." His hand was shaking badly as he took out a checkbook and quickly, nervously scribbled. He passed the check to Harvey, and George managed to get a good look at it. His eyes went wide when he saw the amount it was made out for. Five hundred pounds!

"Now look 'ere, Mr. Kent," Harvey blustered. "I don't like the looks of this bloke. I reckon—"

"I understand your concern," the older man told him, "but I'm sure Mr.—?" He turned to George.

"Er, Smith," George told him, unwilling to reveal his real name. This could still be some crazy joke, but if so, it would be on some bloke called "Smith," and not on George Benson! "I'm sure that Mr. Smith is legitimate. And in any case I daren't miss the chance to get . . . to get back home." He was eager now to be on his way. "Are you ready, Mr. Smith?"

"Just as soon as you say," George told him. "The sooner the better."

They walked out into the night, to George's car, leaving fat greasy Harvey worriedly squeezing his hands in the doorway to his all-nighter. Suddenly the night air seemed inordinately cold, and as George opened the passenger door to let Kent get in, he shivered. He walked around the car, climbed into the driver's seat, and slammed the door.

As George started up the motor, Kent spoke up from where he crouched against the opposite door, a huddled shape in the dark interior of the car. "Are you sure that— that—"

"Look," George answered, the utter craziness of the whole business abruptly dawning on him, souring his voice, "if this is some sort of nutty joke . . ." He let the threat hang, then snapped, "Of course I can find it again. High House, you're talking about?"

"Yes, yes. High House. The home I built for the woman who lives there, waiting for me."

"For fifteen years?" George allowed himself to indulge in the other's fantasy.

"She would wait until time froze!" Kent leaned over to spit the words in George's ear. "And in any case, I have a theory."

*Yeah!* George thought to himself. *Me, too!* Out loud he said, "A theory?"

"Yes. I think—I hope—it's possible that time itself *is*

frozen at the moment of the fracture. If I can get back, it may all be unchanged. I may even regain my lost years!"

"A parallel dimension, eh?" George said, feeling strangely nervous.

"Right." His passenger nodded emphatically. "That's the way I see it."

Humoring him, George asked, "What's it like, this other world of yours?"

"Why, it's just like this world—except that there's a village called Middle Hamborough, and a house on a hill, and a building firm called Milton, Jones & Kent. There are probably other differences, too, but I haven't found any yet to concern me. Do you know the theory of parallel worlds?"

"I've read some science fiction," George guardedly answered. "Some of these other dimensions, or whatever they're supposed to be, are just like this world. Maybe a few odd differences, like you say. Others are different, completely different. Horrible and alien—stuff like that." He suddenly felt stupid. "That's what I've read, anyway. Load of rubbish!"

"Rubbish?" Kent grunted, stirring in his seat. "I wish it were. But anyway, you've got the right idea. Why are you stopping?"

"See that sign?" George said, pointing through the windshield to where the headlights lit up a village name board. "Meadington, just a few miles down the road. We're through Meadington in about five minutes. Then we turn left where it's sign posted Middle Hamborough. Another five minutes after that and we're at High House. You said it would be worth my while?"

*Now comes the crunch,* George told himself. *This is where the idiot bursts out laughing—and that's when I brain him!*

But Kent didn't laugh. Instead he got out his checkbook,

and George switched on the interior light to watch him write a note for . . .

George's eyes bulged as he saw the numbers go down on the crisp paper. First a one, followed by three zeros! One thousand pounds! "This won't bounce?" he asked suspiciously, his hand trembling as he reached for the check.

"It won't bounce," said Kent, folding the note and tucking it into his pocket. "Fortunately, my money was good for this world, too. You get it when we get to High House."

"You have a deal," George told him, putting the car in gear. They drove through slumbering Meadington, its roofs and hedges silvered in a moonlight that shone through the promise of a mist. Leaving the village behind, the car sped along the country road, but after a few minutes George again pulled into the curb and stopped. His passenger had slumped down in his seat.

"Are you OK?" George asked.

"There's no turnoff," Kent sobbed. "We should have passed it before now. I've driven down this road a thousand, ten thousand times in the last fifteen years, and tonight it's just the same as it always is. There's no turnoff, no signpost to Middle Hamborough!"

"Yeah." George chewed his lip, unwilling to accept defeat so easily. "We must have missed it. It wasn't this far out of Meadington last time." He turned the big car about, driving onto the grass verge to do so, then headed back toward Meadington.

George was angry now and more than a little puzzled. He'd been watching for that signpost as keenly as his passenger. How the hell could they have missed it? No matter, this time he'd drive dead slow. He knew the road was there, for he'd been down it and back once already tonight!

Sure enough, with the first of Meadington's roofs glimmering silver in the near distance a dilapidated signpost suddenly showed up in the beam of the car's lights. It pointed across the tarmac to where the surface of a second road ribboned away into the milky moonlight, a sign whose legend, though grimy, was nevertheless amply legible: MIDDLE HAMBOROUGH.

And quite as suddenly George Benson's passenger was sitting bolt upright in his seat, his whole body visibly trembling while his eyes stood out like organ stops, staring madly at the signpost. "Middle Hamborough!" he cried, his voice pitched so high it almost broke. And again: "Middle Hamborough, Middle Hamborough!"

"Sure," said George, an unnatural chill racing up his spine. "I told you I could find it!" And to himself he added, *But I'm damned if I know how we missed it the first time!*

He turned onto the new road, noticing the second signpost at his right as he did so. That was the one they'd missed. Perhaps it had been in the shadows; but in any case, what odds? They were on the right road now.

Kent's trembling had stopped, and his voice was quite steady when he said, "You really don't know how much I owe you, Mr. Smith. You shall have your check, of course, but if it were for a million pounds, it wouldn't really be enough." His face was dark in the car's interior, and his silhouette looked different somehow.

George said, "You realize that fat Harvey's been having you on all this time, don't you?" His voice became quite gentle as he added, "You know, you really ought to see someone about it, about all . . . *this,* I mean. People can take advantage of you. Harvey could have brought you here anytime he wanted."

Suddenly Kent laughed, a young laugh that had the merest trace of weary hysteria in it. "Oh, you don't know the half of it, do you, Mr. Smith? Can't you get it through your head that I'm not mad and no one is trying to make a fool of you? This is all real. My story is the truth. I *was* lost in an alien dimension, in your world, but now I'm finally back in my own. You may believe me, Mr. Smith, that you have earned your thousand pounds!"

George was almost convinced. Certainly Kent was sincere enough. "Well, OK—whatever you say. But I'll tell you something, Mr. Kent. If that check of yours bounces when I try to cash it tomorrow, I'll be back, and you better believe I'll find High House again!"

The silhouette turned in its seat in an attitude of concern. "Do me a favor, will you, Mr. Smith? If—just *if,* you understand—if you can't find the road back to Meadington, don't hesitate to—"

George cut him off with a short bark of a laugh. "You must be joking! I'll find it, all right." His voice went hard again. "And I'll find you, too, if—"

But he paused as, at the top of the next low hill, the headlights illuminated a house standing above the road at the end of a winding drive. George's passenger suddenly gripped his elbow in terrific excitement. "High House!" Kent cried, his voice wild and exultant. "High House! You've done it!"

George grunted in answer, revving the car down into the valley and up the hill to pull it to a halt outside the wrought-iron gates. He reached across to catch hold of his passenger's coat as Kent tried to scramble from the car. "Kent!"

"Oh, yes, your check," said the young man, turning to smile at George in the yellow light from the little lamp on the gate. . . .

George's jaw dropped. Oh, this was Kent, all right. Little doubt about that. Same features, same suit (though it hung a little baggily on him now), same trembling hand that reached into a pocket to bring out the folded check and place it in George's suddenly clammy hand. But it was a hand that trembled now in excitement and not frustrated but undying hope, *and it was a Kent fifteen years younger.*

One thousand pounds, and at last George knew that he had indeed earned it!

Kent turned and threw open the gates, racing up the drive like a wild man. In the house, lights were starting to go on. George fingered the check unbelievingly and ran his tongue over dry lips. His mind seemed to have frozen over, so that only one phrase kept repeating in his brain. It was something Kent had said: "If you can't find the road back to—"

He gunned the motor, spinning the car wildly around in a spray of gravel. Up on the hill at the top of the drive, Kent was vaulting the fence, and a figure in white was waiting in the garden for him, open arms held wide. George tore his eyes away from them and roared down the hill, as for the second time that night he headed for the Meadington road.

The check lay on the empty passenger seat now where he'd dropped it, and money was quite the last thing in George's mind as he drove his car in an unreasoning panic, leaping the low hills like some demon hurdler as he tried to make it back to the main road before—before what? A hideous doubt was blossoming in his mind, growing like some evil genie from a bottle and taking on a horrible form. All those stories about queer dislocations of space and time—the signpost for Middle Hamborough that was, then wasn't, then was again—and, of course, Kent's story, and his . . . rejuvenation?

"I will be very glad," George told himself out loud, "when I reach that junction just outside of Meadington!" For one thing, he could have sworn that it wasn't this much of a drive. He should surely have been there by now. Ah, yes, this would be it coming up now, just around this slight bend. . . .

*No junction!*

The road stretched straight on ahead, narrow and suddenly ominous in the sweeping beam of his lights. All right, so the junction was a little farther than he'd reckoned. George put his foot down even harder to send the big car racing along the narrow road. The miles flew by without a single signpost or junction, and a ground mist came in that forced George to slow down. He would have done so anyway, for now the road seemed to be exerting a strange pull on his car. The big motor felt as if it were slowing down! George's heart almost leapt into his mouth. There couldn't be anything wrong with the car, could there?

Braking to a halt and switching off the car's engine and lights, George climbed out of the driver's seat. He breathed the damp night air. On unpleasantly rubbery legs he walked around to the front of the car and lifted the hood. An inspection light came on and he cast a quick, practiced glance over the motor. No, he'd worked in a garage for many years and he knew a good motor when he saw one. Nothing wrong with the car, so—

As he straightened up, George felt an unaccustomed suction on his shoes and glanced down at the road. The surface was rubbery, formed of a sort of tough sponge. A worried frown crossed George's face as he bent to feel that peculiar surface. He'd never seen a road surfaced with stuff like that before!

It was as he straightened up again that he heard the tinkling, like the sound of tiny bells from somewhere off the road. Yes, there set back from the road, he could see a row of low squat houses, like great mushrooms partly obscured by the mist that swirled now in strange currents. The tinkling came from the houses.

The outskirts of a village? George wondered. Well, at least he'd be able to get directions. He stepped off the road onto turf and made for the houses, only slowing down when he saw how featureless and alike they all looked. The queer tinkling went on, sounding like the gentle noises the hangings on a Christmas tree make in a draft. Other than that there was only the billowing mist and the darkness.

Reaching the first house, stepping very slowly now, George came up close to the wall and stared at it. It was grey, featureless. All the houses looked alike. They were indeed like enormous mushrooms. No windows. Overhanging roofs. Flaps of sorts that could be doors, or there again—

The tinkling had stopped. Very carefully George reached out and touched the wall in front of him. It felt warm . . . and it crept beneath his fingers!

Deliberately and slowly George turned about and forced one foot out in front of the other. He fought the urge to look back over his shoulder until, halfway to the mist-wreathed car, he heard an odd plopping sound behind him. It was like the *ploop* you get throwing a handful of mud into a pond. He froze with his back still to the houses.

Quite suddenly he felt sure that his ears were enlarging, stretching back and up to form saucerlike receivers on top of his head. Everything he had went into those ears, and all of it was trying to tune in on what was going on behind him. He didn't turn, but simply stood still; and again there was only the utter silence, loud in his strangely sensitized ears. He forced his dead feet to take a few more paces forward— and sure enough the sound came again, repeating this time: *ploop, ploop, ploop!*

George slowly pivoted on his heel as muscles he never knew he had began to jump in his face. The noises, each *ploop* sounding closer than the last, stopped immediately. His legs felt like twin columns of jelly, but he somehow completed his turn. He stumbled spastically then, arms flailing to keep himself from falling. The nearest house, or cottage, or whatever, *was right there behind him, within arm's reach!*

Suddenly George's heart, which he was sure had stopped

forever, became audible again inside him, banging away in his chest like a trip hammer. All in one movement he turned and bounded for the car, wondering why with each leap he should stay so long in the air, knowing that in fact his body was moving like greased lightning while his mind (in an even greater hurry, one his body couldn't even attempt to match) thought he was in reverse!

Not bothering, not *daring* to look back again, he almost wrenched the car door from its hinges as he threw himself into the driving seat. Then, in an instant that lasted several centuries, his hand was on the ignition key and the engine was roaring. As he spun the car about in a squeal of tortured tires and accelerated up the rubbery road, he looked in his rearview mirror—and immediately wished he hadn't!

The "houses" were all *ploop*ing down the road after him—like great greedy frogs—and their "doors" were wide open!

George nearly went off the road then, wrenching at the wheel with clammy hands as he fought to control his careening car on the peculiar surface. A million monstrous thoughts raced through his head as he climbed up through the gears. For of course he knew now for certain that he was trapped in an alien dimension, that the space-time elastic had snapped back into place behind him, stranding him here. Wherever "here" was!

It was only several miles later that he thought to slow down, and only then after passing a junction on the right and a signpost saying: MIDDLE HAMBOROUGH 5½ MILES. His heart gave a wild leap as he skidded to a halt on a once more perfectly normal tarmac road. Why, that sign meant that just half a mile up the road in front he'd find Meadington, and beyond Meadington . . . Bankhead and the M1!

Except that Meadington wasn't there. . . . Instead, the mist came up again and, worse, the road went rubbery. And no sign of Meadington. When he saw a row of mushroom "houses" standing back from the road, George did an immediate, violent about-turn, rocking the car dangerously on the rubber road. Trouble with this weird surface was that it gave too much damn traction.

Amazing that he could still think such mundane thoughts in a situation like this. And yet, through all this protracted nightmare, a ray of hope still shone. The road to Middle Hamborough!

Back there, down that road, there was a house on a hill and beyond that a real, if slightly different, world. A world where at least two of the inhabitants owed him a break. From what Kent had told him, it seemed to George that the other world wasn't much different from his own. He could make a go of things there. He gunned his motor back down the road and out of the mist, back onto a decent tarmac surface and into normally dark night, turning left at the leaning signpost onto the now familiar road to Middle Hamborough.

Or was it familiar?

The hedges bordering the road were different somehow, taller, hiding the fields beyond them from the car's probing headlights, and the road seemed narrower than George remembered it. But that must be his imagination acting up after the terrific shocks of the last ten minutes; it *had* to be, for this was the road to Middle Hamborough. Then, cresting the next hill, suddenly George felt that hellish drag on his tires, and his headlights began to do battle with a thickening, swirling mist. At the same time he saw the house atop the next hill, the house set back off the road at the head of a long winding drive. High House!

There were no lights on in the place now, but it was George's refuge nonetheless. Hadn't Kent told him to come back here if he couldn't find his way back to Meadington? George gave a whoop of relief as he swept down into the shallow valley and up the hill toward the wrought-iron roadside gates. They were still open, as Kent had left them; and as he slowed down fractionally, George swung the wheel to the left, turning his car in through the gates. They weren't quite open all the way, though, so that the front of the car slammed them back on their hinges.

Up the drive the front lights of the house instantly came on; two of them that glowed yellow as though shutters had been quickly opened—or lids lifted! George had no time to note anything else—except perhaps that the drive was very white, not the white of gravel but more of leprous flesh—for at that point the car simply stopped as if it had run head-on into a brick wall! George wasn't belted in. He rose up over the steering wheel and crashed through the windshield, automatically turning his shoulder to the glass.

He hit the drive in a shower of glass fragments, screaming and expecting the impact to hurt. It didn't, and then George knew why the car had stopped like that: the drive was as soft and sticky as hot toffee! And it wasn't a drive!

Behind George the wide fleshy ribbon *tasted* the car and, rising up, flicked it easily to one side. Then it tasted George. He had time to scream, barely, and time for one more quite mundane thought—that this wasn't where Kent lived—before that great white chameleon tongue slithered him up the hill to the house, whose entire front below the yellow windows opened up to receive him.

Shortly thereafter the lights went slowly out again, as if someone had lowered shutters, or as if lids had fallen. . . .

# The Pit-Yakker

*I was born on the northeast coast of England a little way
north of the location for "Fruiting Bodies." (Actually, I mod-
eled Garth Bentham pretty closely after my Old Man.) This
next story is set in just such a village as I came up in, and the
"beach" is in just such a state of repair (?) as I report. It's
rumored the Greens are going to get it cleaned up. Well, and
I hope they do. Except . . . sometimes I wonder what they'll
find under all that pit filth.*

When I was sixteen, my father used to say to me:
"Watch what you're doing with the girls; you're an
idiot to smoke, for it's expensive and unhealthy; stay away
from Raymond Maddison!" My mother had died two years
earlier, so he'd taken over her share of the nagging, too.

The girls? Watch what I was doing? At sixteen I barely
*knew* what I was doing! I knew what I wanted to do, but the
how of it was a different matter entirely. Cigarettes? I en-
joyed them; at the five-a-day stage, they still gave me that
occasionally sweet taste and made my head spin. Raymond
Maddison? I had gone to school with him, and because he
lived so close to us we'd used to walk home together. But
his mother was a little weak-minded, his older brother had

been put away for molesting or something, and Raymond himself was thick as two short planks, hulking and unlovely, and a very shadowy character in general. Or at least he gave that impression.

Girls didn't like him: he smelled of bread and dripping and didn't clean his teeth too well, and for two years now he'd been wearing the same jacket and trousers, which had grown pretty tight on him. His short hair and little piggy eyes made him look bristly, and there was that looseness about his lips that you find in certain idiots. If you were told that ladies' underwear was disappearing from washing lines, you'd perhaps think of Raymond. If someone was jumping out on small girls at dusk and shouting *boo!,* he was the one who'd spring to mind. If the little-boy-up-the-road's kitten got strangled . . .

Not that that sort of thing happened a lot in Harden, for it didn't. Up there on the northeast coast in those days, the Bobbies on the beat were still Bobbies, unhampered by modern "ethics" and other humane restrictions. Catch a kid drawing red, hairy, diamond-shaped designs on the school wall, and *wallop!,* he'd get a clout round the ear hole, dragged off home to his parents, and doubtless another wallop. Also, in the schools, the cane was still in force. Young people were still being "brought up," were made or at least encouraged to grow up straight and strong, and not allowed to bolt and run wild. Most of them, anyway. But it wasn't easy, not in that environment.

Harden lay well outside the fringes of "Geordie-land"— Newcastle and environs—but real outsiders termed us all Geordies anyway. It was the way we spoke; our near-Geordie accents leapt between soft and harsh as readily as the Welsh tongue soars up and down the scales; a dialect that at once identified us as "pit-yakkers," grimy-black shambling colliers, coal miners. The fact that my father was a Harden greengrocer made no difference: I came from the

colliery and so was a pit-yakker. I was an apprentice wood-
cutting machinist in Hartlepool?—so what? My collar was
grimy, wasn't it? With coal dust? And no matter how much
I tried to disguise it, I had that accent, didn't I? Pit-yakker!
But at sixteen I *was* escaping from the image. One must,
or sex remain forever a mystery. The girls—the better girls,
anyway—in the big towns, even in Harden, Easington,
Blackhill and the other colliery villages, weren't much im-
pressed by or interested in pit-yakkers. Which must have
left Raymond Maddison in an entirely hopeless position.
Everything about him literally shrieked of his origin, made
worse by the fact that his father, a miner, was already
grooming Raymond for the mine, too. You think I have a
down on them, the colliers? No, for they were the salt of the
earth. They still are. I merely give you the background.

As for my own opinion of Raymond: I thought I knew
him and didn't for a moment consider him a bad sort. He
loved John Wayne like I did, and liked to think of himself
as a tough egg, as I did. But nature and the world in general
hadn't been so kind to him, and being a bit of a dunce didn't
help much either. He was like a big scruffy dog who sits at
the corner of the street grinning at everyone going by and
wagging his tail, whom nobody ever pats for fear of fleas or
mange or whatever, and who you're sure pees on the front
wheel of your car everytime you park it there. He probably
doesn't, but somebody has to take the blame. That was how
I saw Raymond.

So I was sixteen and some months, and Raymond Maddison about the same, and it was a Saturday in July. Normally when we met we'd pass the time of day. Just a few
words: what was on at the cinema (in Harden there were
two of them, the Ritz and the Empress—for this was before
Bingo closed most of them down), when was the next dance
at the Old Victoria Hall, how many pints we'd downed last
Friday at the British Legion. Dancing, drinking, smoking,

girls: it was a time of experimentation. Life had so many flavors other than those that wafted out from the pit and the coke ovens. On this Saturday, however, he was the last person I wanted to see, and the very last I wanted to be seen with.

I was waiting for Moira, sitting on the recreation-ground wall where the stumps of the old iron railings showed through, which they'd taken away thirteen years earlier for the war effort and never replaced. I had been a baby then but it was one of the memories I had: of the men in the helmets with the glass faceplates cutting down all the iron things to melt for the war. It had left only the low wall, which was ideal to sit on. In the summer the flat-capped miners would sit there to watch the kids flying kites in the recreation ground or playing on the swings, or just to sit and talk. There was a group of old-timers there that Saturday, too, all looking out across the dark, fuming colliery toward the sea; so when I saw Raymond hunching my way with his hands in his pockets, I turned and looked in the same direction, hoping he wouldn't notice me. But he already had.

"Hi, Joshua!" he said in his mumbling fashion, touching my arm. I don't know why I was christened "Joshua": I wasn't Jewish or a Catholic or anything. I *do* know why; my father told me *his* father had been called Joshua, so that was it. Usually they called me Josh, which I liked because it sounded like a wild-western name. I could imagine John Wayne being called Josh. But Raymond occasionally forgot and called me "Joshua."

"Hello, Ray*mond!*" I said. I usually called him "Ray," but if he noticed the difference he didn't say anything.

"Game of snooker?" It was an invitation.

"No." I shook my head. "I'm, er, waiting for someone."

"Who?"

"Mind your own business."

"Girl?" he said. "Moira? Saw you with her at the Ritz. Back row."

"Look, Ray, I—"

"It's OK," he said, sitting down beside me on the wall. We're jus' talking. I can go any time."

I groaned inside. He was bound to follow us. He did stupid things like that. I decided to make the best of it, glanced at him. "So, what are you doing? Have you found a job yet?"

He pulled a face. "Naw."

"Are you going to?"

"Pit. Next spring. My dad says."

"Uh-huh." I nodded. "Plenty of work there." I looked along the wall past the groundkeeper's house. That's the way Moira would come.

"Hey, look!" said Raymond. He took out a brand new Swiss Army penknife and handed it over for my inspection. As my eyes widened he beamed. "Beauty, eh?"

And it was. "Where'd you get it?" I asked him, opening it up. It was fitted with every sort of blade and attachment you could imagine. Three or four years earlier I would have loved a knife like that. But right now I couldn't see why I'd need it. OK for wood carving or the Boy Scouts, or even the Boys' Brigade, but I'd left all that stuff behind. And anyway, the machines I was learning to use in my trade paled this thing to insignificance and made it look like a very primitive toy. Like a rasp beside a circular saw. I couldn't see why Raymond would want it either.

"Saved up for it," he said. "See, a saw. Two saws! One for metal, one for wood. Knives—*careful!*—sharp. Gouge—"

"That's an auger," I said, "not a gouge. But . . . this one's a gouge, right enough. Look," and I eased the tool from its housing to show him.

"Corkscrew," he went on. "Scissors, file, hook . . ."

"Hook?"

"For hooking things. Magnetic. You can pick up screws."

"It's a good knife," I told him, giving it back. "How do you use it?"

"I haven't," he said, "—yet."

I was getting desperate. "Ray, do me a favor. Look, I have to stay here and wait for her. And I'm short of cigs." I forked out a florin. "Bring me a packet, will you? Twenty? And I'll give you a few."

He took the coin. "You'll be here?"

I nodded, lying without saying anything. I had an unopened packet of twenty in my pocket. He said no more but loped off across the road, disappearing into one of the back streets leading to Harden's main road and shopping area. I let him get out of sight, then set off briskly past the groundkeeper's house, heading north.

Now, I know I've stated that in my opinion he was OK; but even so, still I knew he wasn't to be trusted. He just *might* follow us, if he could—out of curiosity, perversity, don't ask me. You just couldn't be sure what he was thinking, that's all. And I didn't want him peeping on us.

It dawns on me now that in his "innocence" Raymond was anything but innocent. There are two sides to each of us, and in someone like him, a little lacking in basic understanding . . . well, who is to say that the dark side shouldn't on occasion be just a shade darker? For illustration, there'd been that time when we were, oh, nine or ten years old? I had two white mice who lived in their box in the garden shed. They had their own swimming pool, too, made out of an old baking tray just two and a half inches deep. I'd trained them to swim to a floating tin lid for bits of bacon rind.

One day, playing with Raymond and the mice in the garden, I'd been called indoors about something or other.

I was only inside a moment or two, but when I came back out he'd gone. Looking over the garden wall and down the street, I'd seen him *tip-toeing* off into the distance! A great hulk like him, slinking off like a cartoon cat!

Then I'd shrugged and returned to my game—and just in time. The tin-lid raft was upside down, with Peter and Pan trapped underneath, paddling for all they were worth to keep their snouts up in the air trapped under there with them. It was only a small thing, I suppose, but it had given me bad dreams for a long time. So . . . instead of the hard nut I considered myself, maybe I was just a big softy after all. In some things.

But . . . did Raymond do it deliberately or was it an accident? And if the latter, then why was he slinking off like that? If he had tried to drown them, why? Jealousy? Something I had that he didn't have? Or sheer, downright nastiness? When I'd later tackled him about it, he'd just said: "Eh? Eh?" and looked dumb. That's the way it was with him. I could never figure out what went on in there.

Moira lived down by the high colliery wall, beyond which stood vast cones of coal, piled there, waiting to fuel the coke ovens. And as a backdrop to these black foothills, the wheelhouse towers rising like sooty sentinels, coming into view as I hurried through the grimy sunlit streets; a colliery in the summer seems strangely opposed to itself. In one of the towers a massive spoked wheel was spinning even now, raising or lowering a cage in its claustrophobic shaft. Miners, some still in their "pit black," even wearing their helmets and lamps, drew deep on cigarettes as they came away from the place. My father would have said: "As if their lungs aren't suffering enough already!"

I knew the exact route Moira would take from her gritty colliery-street house to the recreation ground, but at each junction in its turn I scanned the streets this way and that, making sure I didn't miss her. By now Raymond would have bought the cigarettes and be on his way back to the wall.

"Hello, Josh!" she said, breathlessly surprised—almost as if she hadn't expected to see me today—appearing like a ray of extra bright sunlight from behind the freshly creosoted fencing of garden allotments. She stood back and looked me up and down. "So, you're all impatient to see me, eh? Or . . . maybe I was late?" She looked at me anxiously.

I had been hurrying and so was breathing heavily. I smiled, wiped my forehead, said: "It's . . . just that there was someone I knew back there, at the recreation ground, and—"

"You didn't want to be seen with me?" She frowned. She was mocking me, but I didn't know it.

"No, not that," I hurriedly denied it, "but—"

And then she laughed and I knew she'd been teasing. "It's all right, Josh," she said. "I understand." She linked my arm. "Where are we going?"

"Walking," I said, turning her into the maze of allotments, trying to control my breathing, my heartbeat.

"I know *that!*" she said. "But where?"

"Down to the beach, and up again in Blackhill?"

"The beach is very dirty. Not very kind to good clothes." She was wearing a short blue skirt, white blouse, a smart white jacket across her arm.

"The beach banks, then," I gulped. "And along the cliff paths to Easington."

"You only want to get me where it's lonely," she said, but with a smile. "All right, then." And a moment later. "May I have a cigarette?"

I brought out my fresh pack and started to open it, but looking nervously around she said: "Not just yet. When we're farther into the allotments." She was six months my junior and lived close by; if someone saw her smoking it was likely to be reported to her father. But a few minutes later we shared a cigarette and she kissed me, blowing smoke into my mouth. I wondered where she'd learned to do that. Also, it took me by surprise—the kiss, I mean. She was impulsive like that.

In retrospect, I suppose Moira was my first love. And they say you never forget the first one. Well, they mean you never forget the first *time*—but I think your first love is the same, even if there's nothing physical. But she was the first one who'd kept me awake at night thinking of her, the first one who made me ache.

She was maybe five feet six or seven, had a heart-shaped face, huge dark come-to-bed eyes that I suspected and hoped hadn't yet kept their promise, a mouth maybe a fraction too wide, so that her face seemed to break open when she laughed, and hair that bounced on her shoulders entirely of its own accord. They didn't have stuff to make it bounce in those days.

Her figure was fully formed and she looked wonderful in a bathing costume, and her legs were long and tapering. Also, I had a thing about teeth, and Moira's were perfect and very, very white. Since meeting her the first time I'd scrubbed the inside of my mouth and my gums raw trying to match the whiteness of her teeth.

Since meeting her . . .

That had been, oh, maybe three months ago. I mean, I'd always known her, or known of her. You can't live all your life in a small colliery village and not know everyone, at least by sight. But when she'd left school and got her first job at a salon in Hartlepool, and we'd started catching the same bus in the morning, that had opened it up for us.

After that there'd been a lot of talk, then the cinema, eventually the beach at Seaton that the debris from the pits hadn't ruined yet, and now we were "going together." It hadn't meant much to me before, that phrase, "going together," but now I understood it. We went places together, and we went well together. I thought so, anyway.

The garden allotments started properly at the end of the colliery wall and sprawled over many acres along the coast road on the northern extreme of the village. The access paths that divided them were dusty, mazy, meandering. But behind the fences people were at work, and they came to and fro along the paths, so that it wasn't really private there. I had returned Moira's kiss, and in several quieter places had tried to draw her closer once or twice.

Invariably she held me at arm's length, saying: "Not here!" And her nervousness made me nervous, too, so that I'd look here and there all about, to make sure we were unobserved. And it was at such a time, glancing back the way we'd come, that I thought I saw a face hastily snatched back around the corner of a fence. The thought didn't occur to me that it might be Raymond. By now I'd quite forgotten about him.

Where the allotments ended the open fields began, gradually declining to a dene and a stream that ran down to the sea. A second cigarette had been smoked down to its tip and discarded by the time we crossed the fields along a hedgerow, and we'd fallen silent where we strolled through the long summer grass. But I was aware of my arm, linked with hers, hugged close against her right breast. And that was a thought that made me dizzy, for through a heady half hour I had actually held that breast in my hand, had known how warm it was, with its hard little tip that felt rough against the parent softness.

Oh, the back row love seats in the local cinema were worthy of an award; whoever designed them deserves an accolade from all the world's lovers. Two people on a single, softly upholstered seat, thigh to thigh and hip to hip, with no ghastly armrest divider, no obstruction to the slow, breathless, tender, and timid first invasion.

In the dark with only the cinema's wall behind us, and the smoky beam from the projector turning all else to pitch, I was *sure* she wasn't aware of my progress with the top button of her blouse, and I considered myself incredibly fortunate to be able to disguise my fumblings with the second of those small obstacles. But after a while, when for all my efforts it appeared I'd get no farther and my frustration was mounting as the tingling seconds ticked by, then she'd gently taken my hand away and effortlessly completed the job for me. She *had* known—which, while it took something of the edge off my triumph, nevertheless increased the frisson to new and previously unexplored heights.

Was I innocent? I don't know. Others, younger by a year, had said they knew everything there was to know. Everything! That was a thought.

But in opening that button and making way for my hand, Moira had invited me in, as it were; cuddled up together there in the back row, my hand had moulded itself to the shape of her breast and learned every contour better than any actor ever memorized his lines. Even now, a week later, I could form my hand into a cup and feel her flesh filling it again. And *desired* to feel her filling it again.

Where the hedgerow met a fence at right angles, we crossed a stile; I was across first and helped Moira down. While I held one hand to steady her, she hitched her short skirt a little to step down from the stile's high platform. It was funny, but I found Moira's legs more fascinating in that skirt than in her bathing costume. And I'd started to notice the heat of my ears—that they were hot quite apart from the heat of the sun, with a sort of internal burning—as we more nearly approached our destination. My destination, anyway, where if her feelings matched mine she'd succumb a little more to my seductions.

As we left the stile to take the path down into the dene and toward the sea cliffs, I glanced back the way we'd come.

I don't know why. It was just that I had a feeling. And back there, across the fields, but hurrying, I thought . . . a figure. Raymond? If it was, and if he were to bother us today of all days . . . I promised myself he'd pay for it with a bloody nose. But on the other hand it could be anybody. Saying nothing of it to Moira, I hurried her through the dene. Cool under the trees, where the sunlight dappled the rough cobbled path, she said:

"What on earth's the hurry, Josh? Are you *that* eager?"

The way I took her up in my arms and kissed her till I reeled must have answered her question for me; but there were voices here and there along the path, and the place echoed like a tunnel. No, I knew where I wanted to take her.

Toward the bottom of the dene, where it narrowed to a bottleneck of woods and water scooped through the beach banks and funneled toward the sea, we turned north across an old wooden bridge over the scummy stream and began climbing toward the cliff paths, open fields, and sand holes that lay between us and Easington Colliery. Up there, in the long grasses of those summer fields, we could be quite alone and Moira would let me make love to her, I hoped. She'd hinted as much, anyway, the last time I walked her home.

Toiling steeply up an earth track, where white sand spilled down from sand holes up ahead, we looked down on the beach—or what had been a beach before the pit-yakkers came—and remembered a time when it was almost completely white from the banks and cliffs to the sea. On a palmy summer day like this the sea should be blue, but it was grey. Its waves broke in a grey froth of scum on a black shore that looked ravaged by cancer—the cancer of the pits.

The landscape down there could be that of an alien planet: the black beach scarred by streamlets of dully glinting slurry gurgling seaward; concentric tidemarks of congealed froth, with the sick, wallowing sea seeming eager to escape from its own vomit; a dozen sea-coal lorries scattered here and there like ticks on a carcass, their crews shoveling pebble-sized nuggets of the wet, filthy black gold in through open tailgates, while other vehicles trundled like lice over the rotting black corpse of a moonscape. Sucked up by the sun, grey mists wreathed the whole scene.

"It's worse than I remembered it," I said. "And you were right: we couldn't have walked down there, not even along the foot of the banks. It's just too filthy! And to think: all of that was pure white sand just, oh—"

"Ten years ago?" she said. "Well, maybe not *pure* white, but it was still a nice beach then, anyway. Yes, I remember. I've seen that beach full of people, the sea bobbing with their heads. My father used to swim there, with me on his chest! I remember it. I can remember things from all the way back to when I was a baby. It's a shame they've done this to it."

"It's actually unsafe," I told her. "There are places they've flagged, where they've put up warning notices. Quicksands of slag and slop and slurry—gritty black sludge from the pits. And just look at that skyline!"

South lay the colliery at Harden, the perimeter of its works coming close to the banks where they rolled down to the sea, with half a dozen of its black spider legs straddling out farther yet. These were the aerial trip dumpers: conveyor belts or ski lifts of slag, endlessly swaying to the rim and tripped there, to tip the refuse of the coke ovens down onto the smoking wasteland of foreshore; and these were, directly, the culprits of all this desolation. Twenty-four hours a day for fifty years they'd crawled on their high cables, between their spindly towers, great buckets of muck depositing the pus of the earth to corrode a coast. And behind this lower intestine of the works lay the greater pulsating mass of the spider itself: the pit, with its wheel towers and soaring black chimneys, its mastaba cooling towers and mausoleum coke ovens. Yellow smoke, grey

and black smoke, belching continuously into the blue sky—
or into a sky that looked blue but was in fact polluted, as
any rainy day would testify, when white washing on garden
lines would turn a streaky grey with the first patter of
raindrops.

On the southern horizon, Blackhill was a spiky smudge
under a grey haze; north, but closer, Easington was the
same. Viewed from this same position at night, the glow of
the coke ovens, the flare-up and gouting orange steam when
white hot coke was hosed down, would turn the entire
region into a scene straight from hell! Satanic mills? They
have nothing on a nest of well-established coal mines by the
sea. . . .

We reached the top of the banks and passed warning
notices telling how from here on they rolled down to sheer
cliffs. When I'd been a child, miners used to clamber down
the banks to the cliff edge, hammer stakes into the earth and
lower themselves on ropes with baskets to collect gull eggs.
Inland, however, the land was flat, where deep grass pasture
roved wild all the way from here to the coast road. There
were a few farms, but that was all.

We walked half a mile along the cliff path until the fields
began to be fenced; where the first true field was split by a
hedgerow inside the fence, there I paused and turned to
Moira. We hadn't seen anyone, hadn't spoken for some
time but I suppose her heart, like mine, had been speeding
up a little. Not from our efforts, for walking here was easy.

"We can climb the fence, cut along the hedgerow," I
suggested, a little breathlessly.

"Why?" Her eyes were wide, naïve, and yet questioning.

I shrugged. "A . . . shortcut to the main road?" But I'd
made it a question, and I knew I shouldn't leave the initia-
tive to her. Gathering my courage, I added: "Also, we'll—"

"Find a bit of privacy?" Her face was flushed.

I climbed the rough three-bar fence; she followed my

example and I helped her down, and knew she'd seen where I could hardly help looking. But she didn't seem to mind. We stayed close to the hedgerow, which was punctuated every twenty-five paces or so with great oaks, and struck inland. It was only when we were away from the fence that I remembered, just before jumping down, that I'd paused a second to scan the land about—and how for a moment I thought I'd seen someone back along the path. Raymond, I wondered? But in any case, he should lose our trail now.

After some two hundred yards there was a lone elder tree growing in the field a little way apart from the hedge, its branches shading the lush grass underneath. I led Moira away from the hedge and into the shade of the elder, and she came unresisting. And there I spread my jacket for her to sit on, and for a minute or two we just sprawled. The grass hid us almost completely in our first private place. Seated, we could just see the topmost twigs of the hedgerow, and of course the bole and spreading canopy of the nearest oak.

Now, I don't intend to go into details. Anyone who was ever young, alone with his girl, will know the details anyway. Let it suffice to say that there were things I wanted, some of which she was willing to give. And some she wasn't. "No," she said. And more positively: *"No!"* when I persisted. But she panted and moaned a little all the same, and her voice was almost desperate, suggesting: "But I can do it for you this way, if you like." Ah, but her hands set me on fire! I burned for her, and she felt the strength of the flame rising in me. "Josh, *no!*" she said again. "What if . . . if . . ."

She looked away from me, froze for a moment—and her mouth fell open. She drew air hissingly and expelled it in a gasp. *"Josh!"* And without pause she was doing up buttons, scrambling to her feet, brushing away wisps of grass from her skirt and blouse.

"Eh?" I said, astonished. "What is it?"

"He saw us!" she gasped. "He saw you—me—like that!" Her voice shook with a mixture of outrage and fear.

"Who?" I said, mouth dry, looking this was and that and seeing no one. "Where?"

"By the oak tree," she said. "Halfway up it. A face, peering out from behind. Someone was watching us."

Someone? Only one someone it could possibly be! But be sure that when I was done with him he'd never peep on anyone again! Flushed and furious I sprinted through the grass for the oak tree. The hedge hid a rotting fence; I went over, through it, came to a panting halt in fragments of brown, broken timber. No sign of anyone. You could hide an army in that long grass. But the fence where it was nailed to the oak bore the scuff marks of booted feet, and the tree's bark was freshly bruised some six feet up the bole.

"You . . . *dog!*" I growled to myself. "God, but I'll *get* you, Raymond Maddison!"

"Josh!" I heard Moira on the other side of the hedge. "Josh, I'm so—ashamed!"

"What?" I called out. "Of what? He won't dare say anything—whoever he is. There are laws against—"

But she was no longer there. Forcing myself through soft wooden jaws and freeing myself from the tangle of the hedge, I saw her hurrying back the way we'd come. "Moira!" I called, but she was already halfway to the three-bar fence. "Moira!" I called again, and then ran after her. By the time I reached the fence she'd climbed it and was starting back along the path.

I finally caught up with her, took her arm. "Moira, we can find some other place. I mean, just because—"

She shook me off, turned on me. "Is that all you want, Josh Peters?" Her face was angry now, eyes flashing. "Well if it is, there are plenty of other girls in Harden who'll be more than happy to . . . to . . ."

"Moira, I—" I shook my head. It wasn't like that. We were going together.

"I thought you liked *me!*" she snapped. "The real me!" My jaw fell open. Why was she talking to me like this? She knew I liked—more than liked—the real Moira. She *was* the real Moira! It was a tiff, brought on by excitement, fear, frustration; we'd never before had to deal with anything like this, and we didn't know how. My emotions were heightened by hers, and now my pride took over. I thrust my jaw out, turned on my heel, and strode rapidly away from her.

"If that's what you think of me," I called back, "—if that's as *much* as you think of me—then maybe this is for the best. . . ."

"Josh?" I heard her small voice behind me. But I didn't answer, didn't look back.

Furious, I hurried, almost trotted back the way we'd come: along the cliff path, scrambling steeply down through the grass-rimmed, crumbling sand pits to the dene. But at the bottom I deliberately turned left and headed for the beach. Dirty? Oh the beach would be dirty—sufficiently dirty so that she surely wouldn't follow me. I didn't want her to. I wanted nothing of her. *Oh, I did, I did!*—but I wouldn't admit it, not even to myself, not then. But if she did try to follow me, it would mean . . . it would mean . . .

Moira, Moira! Did I love her? Possibly, but I couldn't handle the emotion. So many emotions; and inside I was still on fire from what had nearly been, still aching from the retention of fluids my young body had so desired to be rid of. Raymond? Raymond Maddison? By *God,* but I'd bloody *him!* I'd let some of *his* damned fluids out!

"Josh!" I seemed to hear Moira's voice from a long way back, but I could have been mistaken. In any case it didn't slow me down. Time and space flashed by in a blur; I was down onto the beach; I walked south under the cliffs on sand that was still sand, however blackened; I trekked grimy sand dunes up and down, kicking at withered tufts of crabgrass that reminded me of the grey and yellow hairs sprouting from the blemishes of old men. Until finally I had burned something of the anger and frustration out of myself.

Then I turned toward the sea, cut a path between the sickly dunes down to the no-man's-land of black slag and stinking slurry, and found a place to sit on a rock etched by chemical reaction into an anomalous hump. It was one of a line of rocks I remembered from my childhood, reaching out half a mile to the sea, from which the men had crabbed and cast their lines. But none of that now. Beyond where I sat, only the tips of the lifeless, once limpet- and mussel-festooned rocks stuck up above the slurry; a leaning, blackened signpost warned:

**DANGER! QUICKSAND!**
**DO NOT PROCEED**
**BEYOND THIS POINT.**

Quicksand? Quag, certainly, but not sand . . .

I don't know how long I sat there. The sea was advancing and grey gulls wheeled on high, crying on a rising breeze that blew their plaintive voices inland. Scummy waves broke in feathers of grey froth less than one hundred yards down the beach. Down what had been a beach before the invasion of the pit-yakkers. It was summer but down here there were no seasons. Steam curled up from the slag and misted a pitted, alien landscape.

I became lulled by the sound of the birds, the hissing throb of foamy waters, and, strangely, from some little distance away, the periodic clatter of an aerial dumper tilting its buckets and hurling more mineral debris down from on high, creating a mound that the advancing ocean would spread out in a new layer to coat and further contaminate the beach.

I sat there glumly, with my chin like lead in my hands and all of these sounds dull on the periphery of my consciousness, and thought nothing in particular and certainly nothing of any importance. From time to time a gull's cry would sound like Moira's voice, but too shrill, high, frightened, or desperate. She wasn't coming, wouldn't come, and I had lost her. We had lost each other.

I became aware of time trickling by, but again I state: I don't know how long I sat there. An hour? Maybe.

Then something broke through to me. Something other than the voices of the gulls, the waves, the near-distant rain of stony rubble. A new sound? A presence? I looked up, turned my head to scan north along the dead and rotting beach. And I saw him—though as yet he had not seen me.

My eyes narrowed and I felt my brows come together in a frown. Raymond Maddison. The pit-yakker himself. And this was probably as good a place as any, maybe better than most, to teach him a well-deserved lesson. I stood up, and keeping as low a profile as possible made my way round the back of the tarry dunes to where he was standing. In less than two minutes I was there, behind him, creeping up on him where he stood windblown and almost forlorn seeming, staring out to sea. And there I paused.

It seemed his large, rounded shoulders were heaving. Was he crying? Catching his breath? Gulping at the warm, reeking air? Had he been running? Searching for me? Following me as earlier he'd followed us? My feelings hardened against him. It was because he wasn't entirely all there that people tolerated him. But I more than suspected he *was* all there. Not really a dummy, more a scummy.

And I had him trapped. In front of him the rocks receding into pits of black filth, where a second warning notice leaned like a scarecrow on a battlefield, and behind him . . .

only myself behind him. Me and my tightly clenched fists. Then, as I watched, he took something out of his pocket. His new knife, as I saw now. He stared down at it for a moment, then drew back his arm as if to hurl it away from him, out into the black wilderness of quag. But he froze like that, with the knife still in his hand, and I saw that his shoulders had stopped shuddering. He became alert; I guessed that he'd sensed I was there, watching him.

He turned his head and saw me, and his eyes opened wide in a pale, slack face. I'd never seen him so pale. Then he fell to one knee, dipped his knife into the slurry at his feet, commenced wiping at it with a rag of a handkerchief. Caught unawares he was childlike, tending to do meaningless things.

"Raymond," I said, my voice grimmer than I'd intended. "Raymond, I want a word with you!" And he looked for somewhere to run as I advanced on him. But there was nowhere.

"I didn't—" he suddenly blurted. "I didn't—"

"But you did!" I was only a few paces away.

"I . . . I . . ."

"You followed us, peeped on us, messed it all up."

And again he seemed to freeze, while his brain turned over what I'd said to him. Lines creased his brow, vanishing as quickly as they'd come. "What?"

"*What?!*" I shouted, stepping closer still. "You bloody well *know* what! Now Moira and me, we're finished. And it's your fault."

He backed off into the black mire, which at once covered his boots and the cuffs of his too-short trousers. And there he stood, lifting and lowering his feet, which went *glop, glop* with each up-and-down movement. He reminded me of nothing so much as a fly caught on the sticky paper they used at that time. And his mouth kept opening and closing, stupidly, because he had nothing to say and nowhere to run, and he knew I was angry.

Finally he said: "I didn't mean to . . . follow you. But I—" And he reached into a pocket and brought out a packet of cigarettes. "Your cigarettes."

I had known that would be his excuse. "Throw them to me, Ray," I said. For I wasn't about to go stepping in there after him. He tossed me the packet but stayed right where he was, "You may as well come on out," I told him, lighting up, "for you know I'm going to settle with you."

"Josh," he said, still mouthing like a fish. "Josh . . ."

"Yes, Josh, Josh," I told him, nodding. "But you've really done it this time, and we have to have it out."

He still had his knife. He showed it to me, opened the main blade. He took a pace forward out of the slurry and I took a pace back. There was a sick grin on his face. Except . . . he wasn't threatening me. "For you," he said, snapping the blade shut. "I don't . . . don't want it no more." He stepped from the quag onto a flat rock and stood there facing me, not quite within arm's reach. He tossed the knife and I automatically caught it. It weighed heavy in my hand where I clenched my knuckles round it.

"A bribe?" I said. "So that I won't tell what you did? How many friends do you have, Ray? And how many left if I tell what a dirty, sneaky, spying—"

But he was still grinning his sick, nervous grin. "You won't tell." He shook his head. "Not what I seen."

I made a lunging grab for him and the grin slipped from his face. He hopped to a second rock farther out in the liquid slag, teetered there for a moment before finding his balance. And he looked anxiously all about for more stepping-stones, in case I should follow. There were two or three more rocks, all of them deeper into the coal-dust quicksand, but beyond them only a bubbly, oozy black surface streaked with oil and yellow mineral swirls.

Raymond's predicament was a bad one. Not because of me. I would only hit him. Once or twice, depending how long it took to bloody him. But this stuff would murder him. If he fell in. And the black slime was dripping from the bottoms of his trousers, making the surface of his rock slippery. Raymond's balance wasn't much, neither mentally nor physically. He began to slither this way and that, windmilled his arms in an effort to stay put.

"Ray!" I was alarmed. "Come out of there!"

He leapt, desperately, tried to find purchase on the next rock, slipped! His feet shot up in the air and he came down on his back in the quag. The stuff quivered like thick black porridge and put out slow-motion ripples. He flailed his arms, yelping like a dog, as the lower part of his body started to sink. His trousers ballooned with the air in them, but the stuff's suck was strong. Raymond was going down.

Before I could even start to think straight he was in chest deep, the filth inching higher every second. But he'd stopped yelping and had started thinking. Thinking desperate thoughts. "Josh . . . Josh!" he gasped.

I stepped forward ankle deep, got up onto the first rock. I made to jump to the second rock but he stopped me. "No, Josh," he whispered. "Or we'll both go."

"You're sinking," I said, for once as stupid as him.

"Listen," he answered with a gasp. "Up between the dunes, some cable, half-buried. I saw it on my way down here. Tough, 'lectric wire, in the muck. You can pull me out with that."

I remembered. I had seen it, too. Several lengths of discarded cable, buried in the scummy dunes. All my limbs were trembling as I got back to solid ground, setting out up the beach between the dunes. "Josh!" his voice reached out harshly after me. *"Hurry!"* And a moment later: "The first bit of wire you see, that'll do it. . . ."

I hurried, ran, raced. But my heart was pounding, the air

rasping like sandpaper in my lungs. Fear. But . . . I couldn't find the cable. Then—

There was a tall dune, a great heap of black-streaked, slag-crusted sand. A lookout place! I went up it, my feet breaking through the crust, letting rivulets of sand cascade, thrusting myself to the top. Now I could get directions, scan the area all about. Over there, between low humps of diseased sand, I could see what might be a cable: a thin, frozen black snake of the stuff.

But beyond the cable I could see something else: colors, anomalous, strewn in a clump of dead crabgrass.

I tumbled down the side of the great dune, ran for the cable, tore a length free of the sand and muck. I had maybe fifteen, twenty feet of the stuff. Coiling it, I looked back. Raymond was there in the quag, going down black and sticky. But in the other direction—just over there, no more than a dozen loping paces away, hidden in the crabgrass and low humps of sand—something blue and white and . . . and red.

Something about it made my skin prickle. Quickly, I went to see. And I saw . . .

After a while I heard Raymond's voice over the crying of the gulls. "Josh! *Josh!*"

I walked back, the cable looped in my lifeless hands, made my way to where he hung crucified in the quag; his arms formed the cross, palms pressing down on the belching surface, his head thrown back and the slop ringing his throat. And I stood looking at him. He saw me, saw the cable in my limp hands, looked into my eyes. And he knew. He knew I wasn't going to let him have the cable.

Instead I gave him back his terrible knife with all its terrible attachments—which he'd been waiting to use, and which I'd seen no use for—tossing it so that it landed in front of him and splashed a blob of slime into his right eye.

He pleaded with me for a little while then, but there was

no excuse. I sat and smoked, without even remembering lighting my fresh cigarette, until he began to gurgle. The black filth flooded his mouth, nostrils, the circles of his eyes. He went down, his sputtering mouth forming a ring in the muck that slowly filled in when he was gone. Big shiny bubbles came bursting to the surface. . . .

When my cigarette went out I began to cry, and crying staggered back up to the beach between the dunes. To Moira.

Moira. Something I'd had—almost—that he didn't have. That he could never have, except like this. Jealousy, or just sheer evil? And was I any better than him, now? I didn't know then, and I don't know to this day. He was just a pit-yakker, born for the pit. Him and me both, I suppose, but I had been lucky enough to escape it.

And he hadn't. . . .

# The Mirror of Nitocris

*Impossible to deny Lovecraft's influence on this one, even if I wanted to. It's stamped right there into the title. I think my fascination with Nitocris has its origin in the way she invited all her enemies to a banquet in a great hall under the pyramids, and right in the middle of the romp excused herself, locked them all in, and opened the Nile floodgates! My kind of girl! I believe I looted her name—and possibly the above legend—from macabre mentions in Lovecraft's* Imprisoned With the Pharaohs; *but according to Bob Weinberg, the* Weird Tales *man (the man to ask if you have any queries about the "Unique Magazine" as it was in the Good Old Days), there was another raider long before me, who got there first with another* Weird Tales *story called "The Vengeance of Nitocris." The other guy's name was . . . Tennessee Williams?! Or maybe we're borrowers three and there was a real tyrant-Queen Nitocris of Egypt. No doubt some smart aleck is reaching for his pen right now, to drop me a line and put me right.*

*Anyway, "Mirror" was written in the summer of '68 and made its debut in my very first book, and it's still very high on my "favorite stories" list. I hope you'll like it, too. The style may be starting to fade but the frisson lingers on. . . .*

*Hail, The Queen!*
*Bricked up alive,*

*Never more to curse her hive;*
*Walled-up 'neath the pyramid,*

*Where the sand*
*Her secret hid.*
*Buried with her glass*
*that she,*

*At the midnight hour might see*
*Shapes from other spheres called;*

*Alone with them,*
*Entombed, appalled*
*—to death!*

—Justin Geoffrey

Queen Nitocris's Mirror!
I had heard of it, of course—was there ever an occultist who had not?—I had even read of it, in Geoffrey's raving *People of the Monolith,* and knew that it was whispered of in certain dark circles where my presence is abhorred. I knew Alhazred had hinted of its powers in the forbidden *Necronomicon,* and that certain desert tribesmen still make a heathen sign that dates back untold centuries when questioned too closely regards the legends of its origin.

So how was it that some fool auctioneer could stand up there and declare that this was Nitocris's Mirror? How *dare* he?

Yet the glass was from the collection of Bannister Brown-Farley—the explorer-hunter-archeologist who, before his recent disappearance, was a recognized connoisseur of rare and obscure objets d'art—and its appearance was quite as outré as the appearance of an object with its alleged history ought to be. Moreover, was this not the self same auctioneer, fool or otherwise, who had sold me Baron Kant's silver pistol only a year or two before? Not, mind you, that there was a single shred of evidence that the pistol, or the singular ammunition that came with it, had ever really belonged to the witch-hunting baron; the ornately inscribed "K" on the weapon's butt might stand for anything!

But of course, I made my bid for the mirror, and for Bannister Brown-Farley's diary, and got them both. "Sold to Mr., er, it is Mr. de Marigny, isn't it, sir? Thought so!—sold to Mr. Henri-Laurent de Marigny, for . . ." For an abominable sum.

As I hurried home to the grey stone house that has been my home ever since my father sent me out of America, I could not help but wonder at the romantic fool in me that prompts me all too often to spend my pennies on such pretty tomfooleries as these. Obviously an inherited idiosyncrasy that, along with my love of dark mysteries and obscure and antique wonders, was undoubtedly sealed into my personality as a permanent stamp of my world-famous father, the great New Orleans mystic Etienne-Laurent de Marigny.

Yet if the mirror really *was* once the possession of that awful sovereign—why! What a wonderful addition to my collection. I would hang the thing between my bookshelves, in company with Geoffrey, Poe, d'Erlette, and Prinn. For of course the legends and myths I had heard and read of it were *purely* legends and myths, and nothing more; heaven forbid!

With my ever increasing knowledge of night's stranger mysteries I should have known better.

At home I sat for a long time, simply admiring the thing

where it hung on my wall, studying the polished bronze frame with its beautifully moulded serpents and demons, ghouls and efreets; a page straight out of *The Arabian Nights*. And its surface was so perfect that even the late sunlight, striking through my windows, reflected no glare, but a pure beam of light that lit my study in a dream-engendering effulgence.

Nitocris's Mirror!

Nitocris. Now *there* was a woman—or a monster—whichever way one chooses to think of her. A sixth-dynasty queen who ruled her terror-stricken subjects with a will of supernatural iron from her seat at Gizeh—who once invited all her enemies to a feast in a temple below the Nile, and drowned them by opening the water gates—whose mirror allowed her glimpses of the nether-pits where puffed Shoggoths and creatures of the Dark-Spheres carouse and sport in murderous lust and depravity.

Just suppose this was the real thing, the abhorred glass that they placed in her tomb before sealing her up alive; where could Brown-Farley have got hold of it?

Before I knew it, it was nine, and the light had grown so poor that the mirror was no more than a dull golden glow across the room in the shadow of the wall. I put on my study light, in order to read Brown-Farley's diary, and immediately—on picking up that small, flat book, which seemed to fall open automatically at a well-turned page—I became engrossed with the story that began to unfold. It appeared that the writer had been a niggardly man, for the pages were too closely written, in a crabbed hand, from margin to margin and top to bottom, with barely an eighth of an inch between lines. Or perhaps he had written these pages in haste, begrudging the seconds wasted in turning them and therefore determined to turn as few as possible?

The very first word to catch my eye was *Nitocris*!

The diary told of how Brown-Farley had heard it put

about that a certain old Arab had been caught selling items of fabulous antiquity in the markets of Cairo. The man had been gaoled for refusing to tell the authorities whence the treasures had come. Yet every night in his cell he had called such evil things down on the heads of his gaolers that eventually, in fear, they let him go. And he had blessed them in the name of Nitocris! Yet Abu Ben Reis was not one of those tribesmen who swore by her name—or against it! He was not a Gizeh man, nor even one of Cairo's swarthy sons. His home tribe was a band of rovers wandering far to the east, beyond the great desert. Where, then, had he come into contact with Nitocris's name? Who had taught him her foul blessing—or where had he read of it? For through some kink of fate and breeding Abu Ben Reis had an uncommon knack with tongues and languages other than his own.

Just as thirty-five years earlier the inexplicable *possessions* of one Mohammad Hamad had attracted archeologists of the caliber of Herbert E. Winlock to the eventual discovery of the tomb of Thutmosis III's wives, so now did Abu Ben Reis's hinted knowledge of ancient burial grounds—and in particular the grave of the queen of elder horror—suffice to send Brown-Farley to Cairo to seek his fortune.

Apparently he had not gone unadvised; the diary was full of bits and pieces of lore and legend in connection with the ancient queen. Brown-Farley had faithfully copied from Wardle's *Notes on Nitocris;* and in particular the paragraph on her "Magical Mirror": ". . . handed down to their priests by the hideous gods of inner-Earth before the earliest civilizations of the Nile came into existence—a 'gateway' to unknown spheres and worlds of hellish horror in the shape of a mirror. Worshipped, it was, by the pre-Imer Nyahites in Ptathlia at the dawn of Man's domination of the Earth, and eventually enshrined by Nephren-Ka in a black, windowless crypt on the banks of the Shibeli. Side by side, it lay, with the Shining Trapezohedron, and who can say what things might have been reflected in its depths? Even the Haunter of the Dark may have bubbled and blasphemed before it! Stolen, it remained hidden, unseen for centuries in the bat-shrouded labyrinths of Kith, before finally falling into Nitocris's foul clutches. Numerous the enemies she locked away, the mirror as sole company, full knowing that by the next morning the death-cell would be empty save for the sinister, polished glass on the wall. Numerous the vilely chuckled hints she gave of the *features* of those who leered at midnight from out the bronze-barriered gate. But not even Nitocris herself was safe from the horrors locked in the mirror, and at the midnight hour she was wise enough to gaze but fleetingly upon it. . . ."

The midnight hour! Why! It was ten already. Normally I would have been preparing for bed by this time; yet here I was, so involved, now, with the diary that I did not give my bed a second thought. Better, perhaps, if I had. . . .

I read on. Brown-Farley had eventually found Abu Ben Reis and had plied him with liquor and opium until finally he managed to do that which the proper authorities had found impossible. The old Arab gave up his secret—though the book hinted that this knowledge had not been all *that* easy to extract—and the next morning Brown-Farley had taken a little-used camel track into the wastes beyond those pyramids wherein lay Nitocris's *first* burial place.

But from here on there were great gaps in the writing— whole pages having been torn out or obliterated with thick, black strokes, as though the writer had realized that too much was revealed by what he had written—and there were rambling, incoherent paragraphs on the mysteries of death and the lands beyond the grave. Had I not known the explorer to have been such a fanatical antiquarian (his auctioned collection had been unbelievably varied) and were I

not aware that he had delved, prior to his search for Nitocris's second tomb, into many eldritch places and outré
settings, I might have believed the writer mad from the
contents of the diary's last pages. Even in this knowledge I
half believed him mad anyway.

Obviously he had found the last resting place of Nitocris—the scribbled hints and suggestions were all too
plain—but it seemed there had been nothing left worth
removing. Abu Ben Reis had long since plundered all but
the fabled mirror, and it was after Brown-Farley had taken
that last item from the ghoul-haunted tomb that the first of
his real troubles began. From what I could make out from
the now garbled narrative, he had begun to develop a morbid fixation about the mirror, so that by night he kept it
constantly draped!

But it was no good; before I could continue my perusal
of the diary I had to get down my copy of Feery's *Notes on
the Necronomicon*. There was something tickling me, there
at the back of my mind, a memory, something I should
know, something that Alhazred had known and written
about. As I took down Feery's book from my shelves I
came face to face with the mirror. The light in my study was
bright and the night was quite warm—with that oppressive
heaviness of air that is ever the prelude to violent thunderstorms—yet I shuddered strangely as I saw my face reflected in that glass. Just for a moment it had seemed to leer
at me.

I shrugged off the feeling of dread that immediately
sprang up in my inner self and started to look up the section
concerning the mirror. A great clock chimed out the hour
of eleven somewhere in the night and distant lightning lit up
the sky to the west beyond the windows of my room. One
hour to midnight.

Still, my study *is* the most disconcerting place. What with
those eldritch books on my shelves, their aged leather and

ivory spines dully agleam with the reflection of my study
light; and the *thing* I use as a paperweight, which has no
parallel in any sane or ordered universe; and now with the
mirror and diary, I was rapidly developing an attack of the
fidgets unlike any I had ever known before. It was a shock
for me to realize that I was just a little uneasy!

I thumbed through Feery's often fanciful reconstruction
of the *Necronomicon* until I found the relevant passage. The
odds were that Feery had not altered this section at all;
except, perhaps, to somewhat modernize the "mad" Arab's
old-world phraseology. Certainly it read like genuine Alhazred. Yes, there it was. And there, yet again, was that
recurring hint of happenings at midnight:

"... for while the Surface of the Glass is still—even as the
Crystal Pool of Yith-Shesh, even as the Lake of Hali when
the Swimmers are not at the Frothing—and while its Gates
are locked in all the Hours of Day; yet, at the Witching
Hour, One who knows—even One who guesses—may see in
it all the Shades and Shapes of Night and the Pit, wearing
the Visage of Those who saw before. And though the Glass
may lie forgotten forever its Power may not die, and it
should be known:

*That is not dead which can forever lie,*
*And with strange aeons even death may die. ...*"

For many moments I pondered that weird passage and
the even weirder couplet that terminated it; and the minutes
ticked by in a solemn silence hitherto outside my experience
at The Aspens.

It was the distant chime of the half hour that roused me
from my reverie to continue my reading of Brown-Farley's
diary. I purposely put my face away from the mirror and
leaned back in my chair, thoughtfully scanning the pages.
But there were only one or two pages left to read, and as

best I can remember the remainder of that disjointed narrative rambled on in this manner:

"10th. The nightmares on the *London*—all the way out from Alexandria to Liverpool—Christ knows I wish I'd flown. Not a single night's sleep. Appears the so-called 'legends' are not so fanciful as they seemed. Either that or my nerves are going! Possibly it's just the echo of a guilty conscience. If that old fool Abu hadn't been so damned close-mouthed—if he'd been satisfied with the opium and brandy instead of demanding money—and for what, I ask? There was no need for all that rough stuff. And his poxy waffle about 'only wanting to protect me.' Rubbish! The old beggar'd long since cleaned the place out except for the mirror. . . . That damned mirror! Have to get a grip of myself. What state must my nerves be in that I need to cover the thing up at night? Perhaps I've read too much from the *Necronomicon*! I wouldn't be the first fool to fall for that blasted book's hocus-pocus. Alhazred must have been as mad as Nitocris herself. Yet I suppose it's possible that it's all just imagination; there are drugs that can give the same effects, I'm sure. Could it be that the mirror has a hidden mechanism somewhere that releases some toxic powder or other at intervals? But what kind of mechanism would still be working so perfectly after the centuries that glass must have seen? And why always at midnight? Damned funny! And those *dreams*! There *is* one sure way to settle it, of course. I'll give it a few more days and if things get no better, well—we'll have to wait and see.

"13th. That's it, then. Tonight we'll have it out in the open. I mean, what good's a bloody psychiatrist who insists I'm perfectly well when I know I'm ill? That mirror's behind it all! 'Face your problems,' the fool said, 'and if you do they cease to bother you.' That's what I'll do, then tonight.

"13th. Night. There, I've sat myself down and it's eleven already. I'll wait till the stroke of midnight and then I'll take the cover off the glass and we'll see what we'll see. God!

That a man like me should twitch like this! Who'd believe that only a few months ago I was steady as a rock? And all for a bloody mirror. I'll just have a smoke and a glass. That's better. Twenty minutes to go; good—soon be over now—p'raps tonight I'll get a bit of sleep for a change! The way the place goes suddenly quiet, as though the whole house were *waiting* for something to happen. I'm damned glad I sent Johnson home. It'd be no good to let him see me looking like this. What a god-awful state to get oneself into! Five minutes to go. I'm tempted to take the cover off the mirror right now! There—midnight! Now we'll have it!"

And that was all there was.

I read it through again, slowly, wondering what there was in it that so *alarmed* me. And what a coincidence, I thought, reading that last line for the second time; for even as I did so the distant clock, muffled somewhere by the city's mists, chimed out the hour of twelve.

I thank God, now, that he sent that far-off chime to my ears. I am sure it could only have been an act of Providence that caused me to glance round upon hearing it. For that still glass—that mirror that is quiet as the crystal pool of Yith-Shesh all the hours of day—*was still no longer!*

A *thing,* a bubbling blasphemous shape from lunacy's most hellish nightmare, was squeezing its flabby pulp out through the frame of the mirror into my room—*and it wore a face where no face ever should have been.*

I do not recall moving—opening my desk drawer and snatching out that which lay within—yet it seems I must have done so. I remember only the deafening blasts of sound from the bucking, silver-plated revolver in my clammy hand; and above the rattle of sudden thunder, the whine of flying fragments and the shivering of glass as the hell-forged bronze frame buckled and leapt from the wall.

I remember, too, picking up the strangely *twisted* silver

bullets from my Boukhara rug. And then I must have fainted.

The next morning I dropped the shattered fragments of the mirror's glass overboard from the rail of the Thames Ferry and I melted down the frame to a solid blob and buried it deep in my garden. I burned the diary and scattered its ashes to the wind. Finally, I saw my doctor and had him prescribe a sleeping draft for me. I knew I was going to need it.

I have said the thing had a face.

Indeed, atop the glistening, bubbling mass of that hell-dweller's bulk there *was* a face. A *composite* face of which the two halves did not agree!

*For one of them was the immaculately cruel visage of an ancient queen of Egypt; and the other was easily recognizable—from photographs I had seen in the newspapers—as the now anguished and lunatic features of a certain lately vanished explorer!*

# Necros

*Along with "Fruiting Bodies," "The Pit-Yakker," "No Way Home," and "The Thin People," "Necros" found its way into Karl Edward Wagner's annual collection of Year's Best Horror Stories. To my mind, that makes it good enough for inclusion here. The title has nothing to do with my* Necroscope *novels, even though one of the characters might have stepped right out of them. . . .*

## I

An old woman in a faded blue frock and black head-square paused in the shade of Mario's awning and nodded good day. She smiled a gap-toothed smile. A bulky, slouch-shouldered youth in jeans and a stained yellow T-shirt—a slope-headed idiot, probably her grandson—held her hand, drooling vacantly and fidgeting beside her.

Mario nodded good-naturedly, smiled, wrapped a piece of stale *fucaccia* in greaseproof paper, and came from behind the bar to give it to her. She clasped his hand, thanked him, turned to go.

Her attention was suddenly arrested by something she saw across the road. She started, cursed vividly, harshly, and despite my meagre knowledge of Italian I picked up something of the hatred in her tone. "Devil's spawn!" She said it again. "Dog! Swine!" She pointed a shaking hand and finger, said yet again: "Devil's spawn!" before making the two-fingered, double-handed stabbing sign with which the Italians ward off evil. To do this it was first necessary that she drop her salted bread, which the idiot youth at once snatched up.

Then, still mouthing low, guttural imprecations, dragging the shuffling, *fucaccia*-munching cretin behind her, she hurried off along the street and disappeared into an alley. One word that she had repeated over and over again stayed in my mind: *"Necros! Necros!"* Though the word was new to me, I took it for a curse word. The accent she put on it had been poisonous.

I sipped at my Negroni, remained seated at the small circular table beneath Mario's awning, and stared at the object of the crone's distaste. It was a motorcar, a white convertible Rover and this year's model, inching slowly forward in a stream of holiday traffic. And it was worth looking at it only for the girl behind the wheel. The little man in the floppy white hat beside her—well, he was something else, too. But *she* was—just something else.

I caught just a glimpse, sufficient to feel stunned. That was good. I had thought it was something I could never know again: that feeling a man gets looking at a beautiful girl. Not after Linda. And yet—

She was young, say twenty-four or -five, some three or four years my junior. She sat tall at the wheel, slim, raven haired under a white, wide-brimmed summer hat that just missed matching that of her companion, with a complexion cool and creamy enough to pour over peaches. I stood up—yes, to get a better look—and right then the traffic came to a momentary standstill. At that moment, too, she turned her head and looked at me. And if the profile had stunned me . . . well, the full frontal knocked me dead. The girl was simply, classically, beautiful.

Her eyes were of a dark green but very bright, slightly tilted and perfectly oval under straight, thin brows. Her cheeks were high, her lips a red Cupid's bow, her neck long and white against the glowing yellow of her blouse. And her smile—

—Oh, yes, she smiled.

Her glance, at first cool, became curious in a moment, then a little angry, until finally, seeing my confusion—that smile. And as she turned her attention back to the road and followed the stream of traffic out of sight, I saw a blush of color spreading on the creamy surface of her cheek. Then she was gone.

Then, too, I remembered the little man who sat beside her. Actually, I hadn't seen a great deal of him, but what I had seen had given me the creeps. He too had turned his head to stare at me, leaving in my mind's eye an impression of beady bird eyes, sharp and intelligent in the shade of his hat. He had stared at me for only a moment, and then his head had slowly turned away; but even when he no longer looked at me, when he stared straight ahead, it seemed to me I could feel those raven's eyes upon me, and that a query had been written in them.

I believed I could understand it, that look. He must have seen a good many young men staring at him like that—or rather, at the girl. His look had been a threat in answer to my threat—and because he was practiced in it I had certainly felt the more threatened!

I turned to Mario, whose English was excellent. "She has something against expensive cars and rich people?"

"Who?" He busied himself behind his bar.

"The old lady, the woman with the idiot boy."

"Ah!" He nodded. "Mainly against the little man, I suspect."

"Oh?"

"You want another Negroni?"

"OK—and one for yourself—but tell me about this other thing, won't you?"

"If you like—but you're only interested in the girl, yes?" He grinned.

I shrugged. "She's a good-looker. . . ."

"Yes, I saw her." Now he shrugged. "That other thing—just old myths and legends, that's all. Like your English Dracula, eh?"

"Transylvanian Dracula," I corrected him.

"Whatever you like. And Necros: that's the name of the spook, see?"

"Necros is the name of a vampire?"

"A spook, yes."

"And this is a real legend? I mean, historical?"

He made a fifty-fifty face, his hands palms up. "Local, I guess. Ligurian. I remember it from when I was a kid. If I was bad, old Necros sure to come and get me. Today," again the shrug, "it's forgotten."

"Like the bogeyman." I nodded.

"Eh?"

"Nothing. But why did the old girl go on like that?"

Again he shrugged. "Maybe she think that old man Necros, eh? She crazy, you know? Very backward. The whole family."

I was still interested. "How does the legend go?"

"The spook takes the life out of you. You grow old, spook grows young. It's a bargain you make: he gives you something you want, gets what he wants. What he wants is your youth. Except he uses it up quick and needs more. All the time, more youth."

"What kind of bargain is that?" I asked. "What does the victim get out of it?"

"Gets what he wants," said Mario, his brown face cracking into another grin. "In your case the girl, eh? *If* the little man was Necros. . . ."

He got on with his work and I sat there sipping my Negroni. End of conversation. I thought no more about it—until later.

## II

Of course, I should have been in Italy with Linda, but . . . I had kept her "Dear John" for a fortnight before shredding it, getting mindless drunk, and starting in on the process of forgetting. That had been a month ago. The holiday had already been booked and I wasn't about to miss out on my trip to the sun. And so I had come out on my own. It was hot, the swimming was good, life was easy, and the food superb. With just two days left to enjoy it, I told myself it hadn't been bad. But it would have been better with Linda.

Linda . . . She was still on my mind—at the back of it, anyway—later that night as I sat in the bar of my hotel beside an open bougainvillea-decked balcony that looked down on the bay and the seafront lights of the town. And maybe she wasn't all that far back in my mind—maybe she was right there in front—or else I was just plain daydreaming. Whichever, I missed the entry of the lovely lady and her shriveled companion, failing to spot and recognize them until they were taking their seats at a little table just the other side of the balcony's sweep.

This was the closest I'd been to her, and—

Well, first impressions hadn't lied. This girl *was* beautiful. She didn't look quite as young as she'd first seemed—my own age, maybe—but beautiful she certainly was. And the old boy? He must be, could only be, her father. Maybe it sounds like I was a little naïve, but with her looks this lady really didn't need an old man. And if she did need one it didn't have to be *this* one.

By now she'd seen me and my fascination with her must have been obvious. Seeing it she smiled and blushed at one and the same time, and for a moment turned her eyes away—but only for a moment. Fortunately her companion had his back to me or he must have known my feelings at once; for as she looked at me again—fully upon me this time—I could have sworn I read an invitation in her eyes, and in that same moment any bitter vows I may have made melted away completely and were forgotten. God, *please* let him be her father!

For an hour I sat there, drinking a few too many cocktails, eating olives and potato crisps from little bowls on the bar, keeping my eyes off the girl as best I could, if only for common decency's sake. But . . . all the time I worried frantically at the problem of how to introduce myself, and as the minutes ticked by it seemed to me that the most obvious way must also be the best.

But how obvious would it be to the old boy?

And the damnable thing was that the girl hadn't given me another glance since her original—invitation? Had I mistaken that look of hers?—or was she simply waiting for me to make the first move? *God, let him be her father!*

She was sipping martinis, slowly; he drank a rich red wine, in some quantity. I asked a waiter to replenish their glasses and charge it to me. I had already spoken to the bar steward, a swarthy, friendly little chap from the South called Francesco, but he hadn't been able to enlighten me. The pair were not resident, he assured me; but being resident myself I was already pretty sure of that.

Anyway, my drinks were delivered to their table; they looked surprised; the girl put on a perfectly innocent expression, questioned the waiter, nodded in my direction, and gave me a cautious smile, and the old boy turned his head to stare at me. I found myself smiling in return but avoiding his eyes, which were like coals now, sunken deep

in his brown-wrinkled face. Time seemed suspended—if
only for a second—then the girl spoke again to the waiter
and he came across to me.

"Mr. Collins, sir, the gentleman and the young lady
thank you and request that you join them." Which was
everything I had dared hope for—for the moment.

Standing up I suddenly realized how much I'd had to
drink. I willed sobriety on myself and walked across to their
table. They didn't stand up but the little chap said, "Please
sit." His voice was a rustle of dried grass. The waiter was
behind me with a chair. I sat.

"Peter Collins," I said. "How do you do, Mr.—er?—"

"Karpethes," he answered. "Nichos Karpethes. And this
is my wife, Adrienne." Neither one of them had made the
effort to extend their hands, but that didn't dismay me.
Only the fact that they were married dismayed me. He must
be very, very rich, this Nichos Karpethes.

"I'm delighted you invited me over," I said, forcing a
smile, "but I see that I was mistaken. You see, I thought I
heard you speaking English, and I—"

"Thought we were English?" she finished it for me. "A
natural error. Originally I am Armenian, Nichos is Greek,
of course. We do not speak each other's tongue, but we do
both speak English. Are you staying here, Mr. Collins?"

"Er, yes—for one more day and night. Then—" I
shrugged and put on a sad look. "—Back to England, I'm
afraid."

"Afraid?" the old boy whispered. "There is something to
fear in a return to your homeland?"

"Just an expression," I answered. "I meant I'm afraid
that my holiday is coming to an end."

He smiled. It was a strange, wistful sort of smile, wrin-
kling his face up like a little walnut. "But your friends will
be glad to see you again. Your loved ones—?"

I shook my head. "Only a handful of friends—none of

them really close—and no loved ones. I'm a loner, Mr. Karpethes."

"A loner?" His eyes glowed deep in their sockets and his hands began to tremble where they gripped the table's rim. "Mr. Collins, you don't—"

"We understand," she cut him off. "For although we are together, we, too, in our way, are loners. Money has made Nichos lonely, you see? Also, he is not a well man, and time is short. He will not waste what time he has on frivolous friendships. As for myself—people do not understand our being together, Nichos and I. They pry, and I withdraw. And so I, too, am a loner."

There was no accusation in her voice, but still I felt obliged to say: "I certainly didn't intend to pry, Mrs.—"

"Adrienne." She smiled. "Please. No, of course you didn't. I would not want you to think we thought that of you. Anyway I will *tell* you why we are together, and then it will be put aside."

Her husband coughed, seemed to choke, struggled to his feet. I stood up and took his arm. He at once shook me off—with some distaste, I thought—but Adrienne had already signaled to a waiter. "Assist Mr. Karpethes to the gentleman's room," she quickly instructed in very good Italian. "And please help him back to the table when he has recovered."

As he went Karpethes gesticulated, probably tried to say something to me by way of an apology, choked again, and reeled as he allowed the waiter to help him from the room.

"I'm . . . sorry," I said, not knowing what else to say.

"He has attacks." She was cool. "Do not concern yourself. I am used to it."

We sat in silence for a moment. Finally I began: "You were going to tell me—"

"Ah, yes! I had forgotten. It is a symbiosis."

"Oh?"

"Yes. I need the good life he can give me, and he needs . . . my youth? We supply each other's needs." And so, in a way, the old woman with the idiot boy hadn't been wrong after all. A sort of bargain had indeed been struck. Between Karpethes and his wife. As that thought crossed my mind I felt the short hairs at the back of my neck stiffen for a moment. Gooseflesh crawled on my arms. After all, "Nichos" was pretty close to "Necros," and now this youth thing again. Coincidence, of course. And after all, aren't all relationships bargains of sorts? Bargains struck for better or for worse.

"But for how long?" I asked. "I mean, how long will it work for you?"

She shrugged. "I have been provided for. And he will have me all the days of his life."

I coughed, cleared my throat, gave a strained, self-conscious laugh. "And here's me, the nonpryer!"

"No, not at all, I wanted you to know."

"Well," I shrugged, "—but it's been a pretty deep first conversation."

"First? Did you believe that buying me a drink would entitle you to more than one conversation?"

I almost winced. "Actually, I—"

But then she smiled and my world lit up. "You did not need to buy the drinks," she said. "There would have been some other way."

I looked at her inquiringly. "Some other way to—?"

"To find out if we were English or not."

"Oh!"

"Here comes Nichos now." She smiled across the room. "And we must be leaving. He's not well. Tell me, will you be on the beach tomorrow?"

"Oh—yes!" I answered after a moment's hesitation. "I like to swim."

"So do I. Perhaps we can swim out to the raft? . . ."

"I'd like that very much."

Her husband arrived back at the table under his own steam. He looked a little stronger now, not quite so shriveled somehow. He did not sit but gripped the back of his chair with parchment fingers, knuckles white where the skin stretched over old bones. "Mr. Collins," he rustled, "—Adrienne, I'm sorry. . . ."

"There's really no need," I said, rising.

"We really must be going." She also stood. "No, you stay here, er, Peter? It's kind of you, but we can manage. Perhaps we'll see you on the beach." And she helped him to the door of the bar and through it without once looking back.

## III

They weren't staying at my hotel, had simply dropped in for a drink. That was understandable (though I would have preferred to think that she had been looking for me) for *my* hotel was middling tourist class while theirs was something else. They were up on a hill, high on the crest of a Ligurian spur where a smaller, much more exclusive place nestled in Mediterranean pines. A place whose lights spelled money when they shone up there at night, whose music came floating down from a tiny open-air disco like the laughter of high-living elementals of the air. If I was poetic it was because of her. I mean, that beautiful girl and that weary, wrinkled dried-up walnut of an old man. If anything I was sorry for him. And yet in another way I wasn't.

And let's make no pretense about it—if I haven't said it already, let me say it right now—I wanted her. Moreover, there had been that about our conversation, her beach invitation, which told me that she was available.

The thought of it kept me awake half the night. . . .

\*   \*   \*

I was on the beach at 9:00 A.M.—they didn't show until 11:00. When they did, and when she came out of her tiny changing cubicle—

There wasn't a male head on the beach that didn't turn at least twice. Who could blame them? That girl, in *that* costume, would have turned the head of a sphynx. But—there was something, some little nagging thing, different about her. A maturity beyond her years? She held herself like a model, a princess. But who was it for? Karpethes or me?

As for the old man: he was in a crumpled lightweight summer suit and sunshade hat as usual, but he seemed a bit more perky this morning. Unlike myself he'd doubtless had a good night's sleep. While his wife had been changing he had made his way unsteadily across the pebbly beach to my table and sun umbrella, taking the seat directly opposite me; and before his wife could appear he had opened with:

"Good morning, Mr. Collins."

"Good morning," I answered. "Please call me Peter."

"Peter, then." He nodded. He seemed out of breath, either from his stumbling walk over the beach or a certain urgency that I could detect in his movements, his hurried, almost rude "let's get down to it" manner.

"Peter, you said you would be here for one more day?"

"That's right," I answered, for the first time studying him closely where he sat like some strange garden gnome half in the shade of the beach umbrella. "This is my last day."

He was a bundle of dry wood, a pallid prune, a small, umber scarecrow. And his voice, too, was of straw, or autumn leaves blown across a shady path. Only his eyes were alive. "And you said you have no family, few friends, no one to miss you back in England?"

Warning bells rang in my head. Maybe it wasn't so much urgency in him—which usually implies a goal or ambition still to be realized—but eagerness in that the goal was in sight. "That's correct. I am, was, a student doctor. When I get home I shall seek a position. Other than that there's nothing, no one, no ties."

He leaned forward, bird eyes very bright, claw hand reaching across the table, trembling, and—

Her shadow suddenly fell across us as she stood there in that costume. Karpethes jerked back in his chair. His face was working, strange emotions twisting the folds and wrinkles of his flesh into stranger contours. I could feel my heart thumping against my ribs . . . why I couldn't say. I calmed myself, looked up at her, and smiled.

She stood with her back to the sun, which made a dark silhouette of her head and face. But in that blot of darkness her oval eyes were green jewels. "Shall we swim, Peter?"

She turned and ran down the beach, and of course I ran after her. She had a head start and beat me to the water, beat me to the raft, too. It wasn't until I hauled myself up beside her that I thought of Karpethes: how I hadn't even excused myself before plunging after her. But at least the water had cleared my head, bringing me completely awake and aware.

Aware of her incredible body where it stretched, almost touching mine, on the fiber deck of the gently bobbing raft.

I mentioned her husband's line of inquiry, gasping a little for breath as I recovered from the frantic exercise of our race. She, on the other hand, already seemed completely recovered. She carefully arranged her hair about her shoulders like a fan, to dry in the sunlight, before answering.

"Nichos is not really my husband," she finally said, not looking at me. "I am his companion, that's all. I could have told you last night, but . . . there was the chance that you really were curious only about our nationality. As for any 'veiled threats' he might have issued: that is not unusual. He might not have the vitality of younger men, but jealousy is ageless."

"No," I answered, "he didn't threaten—not that I noticed. But jealousy? Knowing I have only one more day to spend here, what has he to fear from me?"

Her shoulders twitched a little, a shrug. She turned her face to me, her lips inches away. Her eyelashes were like silken shutters over green pools, hiding whatever swam in the deeps. "I am young, Peter, and so are you. And you are very attractive, very . . . eager? Holiday romances are not uncommon."

My blood was on fire. "I have very little money," I said. "We are staying at different hotels. He already suspects me. It is impossible."

"What is?" she innocently asked, leaving me at a complete loss.

But then she laughed, tossed back her hair, already dry, dangled her hands and arms in the water. "Where there's a will . . ." she said.

"You know that I want you—" The words spilled out before I could control or change them.

"Oh, yes. And I want you." She said it so simply, and yet suddenly I felt seared. A moth brushing the magnet candle's flame.

I lifted my head, looked toward the beach. Across seventy-five yards of sparkling water the beach umbrellas looked very large and close. Karpethes sat in the shade just as I had last seen him, his face hidden in shadow. But I knew that he watched.

"You can do nothing here," she said, her voice languid— but I noticed now that she, too, seemed short of breath.

"This," I told her with a groan, "is going to kill me!"

She laughed, laughter that sparkled more than the sun on the sea. "I'm sorry," she sobered. "It's unfair of me to laugh. But—your case is not hopeless."

"Oh?"

"Tomorrow morning, early, Nichos has an appointment

with a specialist in Genova. I am to drive him into the city tonight. We'll stay at a hotel overnight."

I groaned my misery. "Then my case *is* quite hopeless. I fly tomorrow."

"But if I sprained my wrist," she said, "and so could not drive . . . and if he went into Genova by taxi while I stayed behind with a headache—because of the pain from my wrist—" Like a flash she was on her feet, the raft tilting, her body diving, striking the water into a spray of diamonds.

Seconds for it all to sink in—and then I was following her, laboring through the water in her churning wake. And as she splashed from the sea, seeing her stumble, go to her hands and knees in Ligurian shingle—and the pained look on her face, the way she held her wrist as she came to her feet. As easy as that!

Karpethes, struggling to rise from his seat, stared at her with his mouth agape. Her face screwed up now as I followed her up the beach. And Adrienne holding her "sprained" wrist and shaking it, her mouth forming an elongated "O." The sinuous motion of her body and limbs, mobile marble with dew of ocean clinging saltily . . .

If the tiny man had said to me: "I am Necros. I want ten years of your life for one night with her," at that moment I might have sealed the bargain. Gladly. But legends are legends and he wasn't Necros, and he didn't, and I didn't. After all, there was no need. . . .

## IV

I suppose my greatest fear was that she might be "having me on," amusing herself at my expense. She was, of course, "safe" with me—insofar as I would be gone tomorrow and the "romance" forgotten, for her, anyway—and I could also see how she was starved for young companionship, a fact she had brought right out in the open from the word go.

But why me? Why should I be so lucky?

Attractive? Was I? I had never thought so. Perhaps it was because I *was* so safe: here today and gone tomorrow, with little or no chance of complications. Yes, that must be it. *If* she wasn't simply making a fool of me. She might be just a tease—

—But she wasn't.

At 8:30 that evening I was in the bar of my hotel—had been there for an hour, careful not to drink too much, unable to eat—when the waiter came to me and said there was a call for me on the reception telephone. I hurried out to reception where the clerk discreetly excused himself and left me alone.

"Peter?" Her voice was a deep well of promise. "He's gone. I've booked us a table, to dine at 9:00. Is that all right for you?"

"A table? Where?" my own voice was breathless.

"Why, up here, of course! Oh, don't worry, it's perfectly safe. And anyway, Nichos knows."

"Knows?" I was taken aback, a little panicked. "What does he know?"

"That we're dining together. In fact he suggested it. He didn't want me to eat alone—and since this is your last night . . ."

"I'll get a taxi right away," I told her.

"Good. I look forward to . . . seeing you. I shall be in the bar."

I replaced the telephone in its cradle, wondering if she always took an apéritif before the main course. . . .

I had smartened myself up. That is to say, I was immaculate. Black bow tie, white evening jacket (courtesy of C & A), black trousers, and a lightly frilled white shirt, the only one I had ever owned. But I might have known that my appearance would never match up to hers. It seemed that everything she did was just perfectly right. I could only hope that that meant literally everything.

But in her black lace evening gown with its plunging neckline, short wide sleeves, and delicate silver embroidery, she was stunning. Sitting with her in the bar, sipping our drinks—for me a large whisky and for her a tall Cinzano—I couldn't take my eyes off her. Twice I reached out for her hand and twice she drew back from me.

"Discreet they may well be," she said, letting her oval green eyes flicker toward the bar, where guests stood and chatted, and back to me, "but there's really no need to give them occasion to gossip."

"I'm sorry, Adrienne," I told her, my voice husky and close to trembling, "but—"

"How is it," she demurely cut me off, "that a good-looking man like you is—how do you say it?—'going short?' "

I sat back, chuckled. "That's a rather unladylike expression," I told her.

"Oh? And what I've planned for tonight is ladylike?"

My voice went huskier still. "Just what is your plan?"

"While we eat," she answered, her voice low, "I shall tell you." At which point a waiter loomed, towel over his arm, inviting us to accompany him to the dining room.

Adrienne's portions were tiny, mine huge. She sipped a slender, light white wine, I gulped blocky rich red from a glass the waiter couldn't seem to leave alone. Mercifully I was hungry—I hadn't eaten all day—else that meal must surely have bloated me out. And all of it ordered in advance, the very best in quality cuisine.

"This," she eventually said, handing me her key, "fits the

door of our suite." We were sitting back, enjoying liqueurs and cigarettes. "The rooms are on the ground floor. Tonight you enter through the door, tomorrow morning you leave via the window. A slow walk down to the seafront will refresh you. How is that for a plan?"

"Unbelievable!"

"You don't believe it?"

"Not my good fortune, no."

"Shall we say that we both have our needs?"

"I think," I said, "that I may be falling in love with you. What if I don't wish to leave in the morning?"

She shrugged, smiled, said: "Who knows what tomorrow may bring?"

How could I ever have thought of her simply as another girl? Or even an ordinary young woman? Girl she certainly was, woman, too, but so . . . *knowing!* Beautiful as a princess and knowing as a whore.

If Mario's old myths and legends were reality, and if Nichos Karpethes were really Necros, then he'd surely picked the right companion. No man born could ever have resisted Adrienne, of that I was quite certain. These thoughts were in my mind—but dimly, at the back of my mind—as I left her smoking in the dining room and followed her directions to the suite of rooms at the rear of the hotel. In the front of my mind were other thoughts, much more vivid and completely erotic.

I found the suite, entered, left the door slightly ajar behind me.

The thing about an Italian room is its size. An entire suite of rooms is vast. As it happened I was only interested in one room, and Adrienne had obligingly left the door to that one open.

I was sweating. And yet . . . I shivered.

Adrienne had said fifteen minutes, time enough for her to smoke another cigarette and finish her drink. Then she would come to me. By now the entire staff of the hotel probably knew I was in here, but this was Italy.

# V

I shivered again. Excitement? Probably.

I threw off my clothes, found my way to the bathroom, took the quickest shower of my life. Drying myself off, I padded back to the bedroom.

Between the main bedroom and the bathroom a smaller door stood ajar. I froze as I reached it, my senses suddenly alert, my ears seeming to stretch themselves into vast receivers to pick up any slightest sound. For there had been a sound, I was sure of it, from that room. . . .

A scratching? A rustle? A whisper? I couldn't say. But a sound, anyway.

Adrienne would be coming soon. Standing outside that door I slowly recommenced toweling myself dry. My naked feet were still firmly rooted, but my hands automatically worked with the towel. It was nerves, only nerves. There had been no sound, or at worst only the night breeze off the sea, whispering in through an open window.

I stopped toweling, took another step toward the main bedroom, heard the sound again. A small, choking rasp. A tiny gasping for air.

Karpethes? What the hell was going on?

I shivered violently, my suddenly chill flesh shuddering in an uncontrollable spasm. But . . . I forced myself to action, returned to the main bedroom, quickly dressed (with the exceptions of my tie and jacket), and crept back to the small room.

Adrienne must be on her way to me even now. She mustn't find me poking my nose into things, like a suspicious kid. I must kill off this silly feeling that had my skin crawling. Not that an attack of nerves was unnatural in the circumstances, on the contrary, but I wasn't about to let it spoil the night. I pushed open the door of the room, entered into darkness, found the light switch. Then—

—I held my breath, flipped the switch.

The room was only half as big as the others. It contained a small single bed, a bedside table, a wardrobe. Nothing more, or at least nothing immediately apparent to my wildly darting eyes. My heart, which was racing, slowed and began to settle toward a steadier beat. The window was open, external shutters closed—but small night sounds were finding their way in through the louvres. The distant sounds of traffic, the toot of horns—holiday sounds from below.

I breathed deeply and gratefully, and saw something projecting from beneath the pillow on the bed. A corner of card or of dark leather, like a wallet or—

—Or a passport!

A Greek passport, Karpethes's, when I opened it. But how could it be? The man in the photograph was young, no older than me. His birth date proved it. And there was his name: Nichos Karpethes. Printed in Greek, of course, but still plain enough. His son?

Puzzling over the passport had served to distract me. My nerves had steadied up. I tossed the passport down, frowned at it where it lay upon the bed, breathed deeply once more . . . and froze solid!

A scratching, a hissing, a dry grunting—from the wardrobe.

Mice? Or did I in fact smell a rat?

Even as the short hairs bristled on the back of my neck I knew anger. There were too many unexplained things here. Too much I didn't understand. And what was it I feared? Old Mario's myths and legends? No, for in my experience the Italians are notorious for getting things wrong. Oh, yes, notorious . . .

I reached out, turned the wardrobe's doorknob, yanked the doors open.

At first I saw nothing of any importance or significance. My eyes didn't know what they sought. Shoes, patent leather, two pairs, stood side by side below. Tiny suits, no bigger than boys' sizes, hung above on steel hangers. And— my God, my God—a waistcoat!

I backed out of that little room on rubber legs, with the silence of the suite shrieking all about me, my eyes bugging, my jaw hanging slack—

"Peter?"

She came in through the suite's main door, came floating toward me, eager, smiling, her green eyes blazing. Then blazing their suspicion, their anger as they saw my condition. "Peter!"

I lurched away as her hands reached for me, those hands I had never yet touched, which had never touched me. Then I was into the main bedroom, snatching my tie and jacket from the bed (don't ask me why!), and out of the window, yelling some inarticulate, choking thing at her and lashing out frenziedly with my foot as she reached after me. Her eyes were bubbling green hells. *"Peter!"*

Her fingers closed on my forearm, bands of steel containing a fierce, hungry heat. And strong as two men she began to lift me back into her lair!

I put my feet against the wall, kicked, came free, and crashed backward into shrubbery. Then up on my feet, gasping for air, running, tumbling, crashing into the night. Down madly tilting slopes, through black chasms of mountain pine with the Mediterranean stars winking overhead, and the beckoning, friendly lights of the village seen occasionally below. . . .

In the morning, looking up at the way I had descended

and remembering the nightmare of my panic flight, I counted myself lucky to have survived it. The place was precipitous. In the end I *had* fallen, but only for a short distance. All in utter darkness, and my head striking something hard. But . . .

I did survive. Survived both Adrienne and my flight from her.

And waking with the dawn, and gently fingering my bruises and the massive bump on my forehead, I made my staggering way back to my still slumbering hotel, let myself in, and *locked* myself in my room—then sat there trembling and moaning until it was time for the coach.

Weak? Maybe I was, maybe I am.

But on my way into Genova, with people round me and the sun hot through the coach's windows, I could think again. I could roll up my sleeve and examine that claw mark of four slim fingers and a thumb, branded white into my suntanned flesh, where hair would never more grow on skin sere and wrinkled.

And seeing those marks I could also remember the wardrobe and the waistcoat—and what the waistcoat contained.

That tiny puppet of a man, alive still but barely, his stick arms dangling through the waistcoat's armholes, his baby's head projecting, its chin supported by the tightly buttoned waistcoat's breast. And the large bulldog clip over the hanger's bar, its teeth fastened in the loose, wrinkled skin of his walnut head, holding it up. And his skinny little legs dangling, twig-things twitching there; and his pleading, pleading eyes!

But eyes are something I mustn't dwell upon.

And green is a color I can no longer bear. . . .

# The Thin People

*For three years I lived in Crouch End, in the north of London.
Do you remember the Stephen King story? I think that one
came out of a visit he paid to Peter Straub, when Straub was
living in Crouch End. We used to have a saying, my wife and
I: "All roads lead to Crouch End." Peter Tremayne lived not
far away (and still does); Clive Barker did one of his plays
up there (those were the early days, before his books started
bleeding); Douglas Hill lived there. Others whose names escape me now . . .*

*Up there in Crouch End, you'd bump into some weird
people. And there were these weird, thin houses. On my way
back from the pub one night, I found myself looking at a thin
house and wondering who in the world could live in such a
cramped-up, concertinaed sort of place. The answer seemed
obvious.*

## I

Funny place, Barrows Hill. Not *Barrow's* Hill, no. Barrows without the apostrophe. For instance: you won't
find it on any map. You'll find maps whose borders approach it, whose corners impinge, however slightly, upon it,

but in general it seems that cartographers avoid it. It's too far out from the center for the tubes, hasn't got a main-line station, has lost much of its integrity by virtue of all the infernal demolition and reconstruction going on around and within it. But it's still there. Buses run to and from, and the older folk who live there still call it Barrows Hill.

When I went to live there in the late '70s I hated the place. There was a sense of senility, of inherent idiocy about it. A damp sort of place. Even under a hot summer sun, damp. You could feel blisters of fungus rising even under the freshest paint. Not that the place got painted very much. Not that I saw, anyway. No, for it was like somewhere out of Lovecraft: decaying, diseased, inbred.

Barrows Hill. I didn't stay long, a few months. Too long, really. It gave you the feeling that if you delayed, if you stood still for just one extra moment, then that it would grow up over you and you'd become a part of it. There are some old, old places in London, and I reckoned Barrows Hill was of the oldest. I also reckoned it for its genius loci; like it was a focal point for secret things. Or perhaps not a focal point, for that might suggest a radiation—a spreading outward—and as I've said, Barrows Hill was ingrown. The last bastion of the strange old things of London. Things like the thin people. The very tall, very thin people.

Now nobody—but nobody *anywhere*—is ever going to believe me about the thin people, which is one of the two reasons I'm not afraid to tell this story. The other is that I don't live there anymore. But when I did . . .

I suspect now, that quite a few people—ordinary people, that is—knew about them. They wouldn't admit it, that's all, and probably still won't. And since all of the ones who'd know live on Barrows Hill, I really can't say I blame 'em. There was one old lad lived there, however, who knew *and* talked about them. To me. Since he had a bit of a reputation (to be frank, they called him "Balmy Bill of Barrows

Hill") I didn't pay a deal of attention at first. I mean, who would?

Barrows Hill had a pub, a couple of pubs, but the one most frequented was The Railway. A hangover from a time when there really was a railway, I supposed. A couple of years ago there had been another, a serious rival to The Railway for a little while, when someone converted an old block into a fairly modern pub. But it didn't last. Whoever owned the place might have known something, but probably not. Or he wouldn't have been so stupid as to call his place The Thin Man! It was only open for a week or two before burning down to the ground.

But that was before my time and the only reason I make mention of pubs, and particularly The Railway, is because that's where I met Balmy Bill. He was there because of his disease, alcoholism, and I was there because of mine, heart-sickness—which, running at a high fever, showed all signs of mutating pretty soon into Bill's problem. In short, I was hitting the bottle.

Now this is all incidental information, of course, and I don't intend to go into it other than to say it was our problems brought us together. As unlikely a friendship as any you might imagine. But Balmy Bill was good at listening, and I was good at buying booze. And so we were good company.

One night, however, when I ran out of money, I made the mistake of inviting him back to my place. (My place—hah! A bed, a loo, and a typewriter; a poky little place up some wooden stairs, like a penthouse kennel; oh, yes, and a bonus in the shape of a cupboard converted to a shower.) But I had a couple bottles of beer back there and a half-bottle of gin, and when I'd finished crying on Balmy Bill's shoulder it wouldn't be far for me to fall into bed. What did surprise me was how hard it was to get him back there. He started complaining the moment we left the bar—or rather, as soon as he saw which way we were headed.

"Up the Larches? You live up there, off Barchington Road? Yes, I remember you told me. Well, and maybe I'll just stay in the pub a while after all. I mean, if you live right up *there*—well, it's out of my way, isn't it?"

"Out of your way? It's a ten-minute walk, that's all! I thought you were thirsty?"

"Thirsty I am—always! Balmy I'm not—they only say I am 'cos they're frightened to listen to me."

"They?"

"People!" he snapped, sounding unaccustomedly sober. Then, as if to change the subject: "A half-bottle of gin, you said?"

"That's right, Gordon's. But if you want to get back on down to The Railway . . ."

"No, no, we're halfway there now," he grumbled, hurrying along beside me, almost taking my arm in his nervousness. "And anyway, it's a nice bright night tonight. They're not much for light nights."

"They?" I asked again.

"People!" Despite his short, bowed legs, he was half a pace ahead of me. "The thin people." But where his first word had been a snarl, his last three were whispered, so that I almost missed them entirely.

Then we were up Larches Avenue—*the* Larches as Balmy Bill had it—and closing fast on twenty-two, and suddenly it was very quiet. Only the scrape of dry, blown leaves on the pavement. Autumn, and the trees half-naked. Moonlight falling through webs of high, black, brittle branches.

"Plenty of moon," said Bill, his voice hushed. "Thank God—in whom I really don't believe—for that. But *no streetlights!* You see that? Bulbs all missing. That's them."

"Them?" I caught his elbow, turning him into my gateway—if there'd been a gate. There wasn't, just the post, which served as my landmark whenever I'd had a skinful.

"Them, yes!" he snapped, staring at me as I turned my key in the lock. "Damn young fool!"

And so up the creaky stairs to my little cave of solitude, and Balmy Bill shivering despite the closeness of the night and warmth of the place, which leeched a lot of its heat from the houses on both sides, and from the flat below, whose elderly lady occupier couldn't seem to live in anything other than an oven; and in through my own door, into the 'living' room, where Bill closed the curtains across the jutting bay windows as if he'd lived there all of his life. But not before he'd peered out into the night street, his eyes darting this way and that, round and bright in his lined, booze-desiccated face.

Balmy, yes. Well, maybe he was and maybe he wasn't.

"Gin," I said, passing him the bottle and a glass. "But go easy, yes? I like a nip myself, you know."

"A nip? A nip? Huh! If I lived here I'd need more than a nip. This is the middle of it, this is. The very middle!"

"Oh?" I grinned. "Myself, I had it figured for the living end!"

He paced the floor for a few moments—three paces there, three back—across the protesting boards of my tiny room, before pointing an almost accusing finger at me. "Chirpy tonight, aren't you? Full of beans!"

"You think so?" Yes, he was right. I did feel a bit brighter. "Maybe I'm over it, eh?"

He sat down beside me. "I certainly hope so, you daft young sod! And now maybe you'll pay some attention to my warnings and get yourself a place well away from here."

"Your warnings? Have you been warning me, then?" It dawned on me that he had, for several weeks, but I'd been too wrapped up in my own misery to pay him much heed. And who would? After all, he was Balmy Bill.

" 'Course I have!" he snapped. "About them bloody—"

"—Thin people," I finished it for him. "Yes, I remember now."

"Well?"

"Eh?"

"Are you or aren't you?"

"I'm listening, yes."

"No, no, *no!* Are you or aren't you going to find yourself new lodgings?"

"When I can afford it, yes."

"You're in danger here, you know. They don't like strangers. Strangers change things, and they're against that. They don't like anything strange, nothing new. They're a dying breed, I fancy, but while they're here they'll keep things the way they like 'em."

"OK," I sighed. "This time I really am listening. You want to start at the beginning?"

He answered my sigh with one of his own, shook his head impatiently. "Daft young bugger! If I didn't like you I wouldn't bother. But all right, for your own good, one last time . . . just listen and I'll tell you what I know. It's not much, but it's the last warning you'll get. . . ."

## II

"Best thing ever happened for 'em must have been the lampposts, I reckon."

"Dogs?" I raised my eyebrows.

He glared at me and jumped to his feet. "Right, that's it, I'm off!"

"Oh, sit down, sit down!" I calmed him. "Here, fill your glass again. And I promise I'll try not to interrupt."

"Lampposts!" he snapped, his brows black as thunder. But he sat and took the drink. "Yes, for they imitate them, see? And thin, they can hide behind them. Why, they can stand so still that on a dark night you wouldn't know the difference! Can you imagine that, eh? Hiding behind or imitating a lamppost!"

I tried to imagine it, but: "Not really," I had to admit. Now, however, my levity was becoming a bit forced. There was something about his intensity—the way his limbs shook in a manner other than alcoholic—that was getting through to me. "Why should they hide?"

"Freaks! Wouldn't you hide? A handful of them, millions of us. We'd hound 'em out, kill 'em off!"

"So why don't we?"

" 'Cos we're all smart young buggers like you, that's why! 'Cos we don't *believe* in 'em."

"But you do."

Bill nodded, his three- or four-day growth of hair quivering on jowls and upper lip. "Seen 'em," he said, "and seen . . . *evidence* of them."

"And they're real people? I mean, you know, human? Just like me and you, except . . . thin?"

"And tall. Oh—*tall!*"

"Tall?" I frowned. "Thin and tall. How tall? Not as tall as—"

"Lampposts." He nodded. "Yes. Not during the day, mind you, only at night. At night they—" (he looked uncomfortable, as if it had suddenly dawned on him how crazy this all must sound) "—they sort of, well, kind of *unfold* themselves."

I thought about it, nodded. "They unfold themselves. Yes, I see."

"No, you don't see." His voice was flat, cold, angry now. "But you will, if you hang around here long enough."

"Where do they live," I asked, "these tall, thin people?"

"In thin houses," he answered, matter-of-factly.

"Thin houses?"

"Sure! Are you telling me you haven't noticed the thin houses? Why, this place of yours very nearly qualifies! Thin houses, yes. Places where normal people wouldn't dream of setting up. There's half a dozen such in Barchington, and a couple right here in the Larches!" He shuddered and I bent to turn on an extra bar in my electric fire.

"Not cold, mate," Bill told me then. "Hell no! Enough booze in me to keep me warm. But I shudder every time I think of 'em. I mean, what *do* they do?"

"Where do they work, you mean?"

"Work?" He shook his head. "No, they don't work. Probably do a bit of tea-leafing. Burglary, you know. Oh, they'd get in anywhere, the thin people. But what do they *do?*"

I shrugged.

"I mean, me and you, we watch telly, play cards, chase the birds, read the paper. But them . . ."

It was on the tip of my tongue to suggest maybe they go into the woods and frighten owls, but suddenly I didn't feel half so flippant. "You said you'd seen them?"

"Seen 'em sure enough, once or twice," he confirmed. "And weird! One, I remember, came out of his thin house— thin house in Barchington, I could show you it sometime in daylight. Me, I was behind a hedge sleeping it off. Don't ask me how I got there, drunk as a lord! Anyway, something woke me up.

"Down at its bottom the hedge was thin where cats come through. It was night and the council men had been round during the day putting blubs in the streetlights, so the place was all lit up. And directly opposite, there's this thin house and its door slowly opening; and out comes this bloke into the night, half of him yellow from the lamplight and half black in shadow. See, right there in front of the thin house is a street lamp.

"But this chap looks normal enough, you know? A bit stiff in his movements: he sort of moves jerky, like them contortionists who hook their feet over their shoulders and walk on their hands. Anyway, he looks up and down the street, and he's obviously satisfied no one's there. Then—

"—He slips back a little into the shadows until he comes up against the wall of his house, and he—unfolds!

"I see the light glinting down one edge of him, see it suddenly split into two edges at the bottom, sort of hinged at the top. And the split widens until he stands in the dark there like a big pair of dividers. And then one half swings up until it forms a straight line, perpendicular—and now he's ten feet tall. Then the same again, only this time the division takes place in the middle. Like . . . like a joiner's wooden three-foot ruler, with hinges so he can open it up, you know?"

I nodded, fascinated despite myself. "And that's how they're built, eh? I mean, well, hinged?"

"Hell, no!" he snorted. "You can fold your arms on their elbows, can't you? Or your legs on their knees? You can bend from the waist and touch your toes? Well I sure can! Their joints may be a little different from ours, that's all— maybe like the joints of certain insects. Or maybe not. I mean, their science is different from ours, too. Perhaps they fold and unfold themselves the same way they do it to other things—except it doesn't do them any harm. I dunno . . ."

"What?" I asked, puzzled. "What other things do they fold?"

"I'll get to that later," he told me darkly, shivering. "Where was I?"

"There he was," I answered, "all fifteen foot of him, standing in the shadows. Then—?"

"A car comes along the street, sudden like!" Bill grabbed my arm.

"Wow!" I jumped. "He's in trouble, right?"

Balmy Bill shook his head. "No way. The car's lights are on full, but that doesn't trouble the thin man. He's not stupid. The car goes by, lighting up the walls with its beam, and where the thin man stood in shadows against the wall of his thin house—"

"Yes?"

"A drainpipe, all black and shiny!"

I sat back. "Pretty smart."

"You better believe they're smart. Then, when it's dark again, out he steps. And *that's* something to see! Those giant strides—but quick, almost a flicker. Blink your eyes and he's moved—and between each movement his legs coming together as he pauses, and nothing to see but a pole. Up to the lamppost he goes, seems almost to melt into it, hides behind it. And *plink!*—out goes the light. After that . . . in ten minutes he had the whole street black as night in a coal mine. And yours truly lying there in somebody's garden, scared and shivering and dying to throw up."

"And that was it?"

Balmy Bill gulped, tossed back his gin, and poured himself another. His eyes were huge now, skin white where it showed through his whiskers. "God, no—that wasn't it—there was more! See, I figured later that I must have got myself drunk deliberately that time—so's to go up there and spy on 'em. Oh, I know that sounds crazy now, but you know what it's like when you're drunk mindless. Jesus, these days I can't *get* drunk! But these were early days I'm telling you about."

"So what happened next?"

"Next—he's coming back down the street! I can hear him: *click,* pause . . . *click,* pause . . . *click,* pause, stilting it along the pavement—and I can see him in my mind's eye, doing his impression of a lamppost with every pause. And suddenly I get this feeling, and I sneak a look around. I mean, the frontage of this garden I'm in is so tiny, and the house behind me is—"

I saw it coming. "Jesus!"

"A thin house," he confirmed it, "right!"

"So now *you* were in trouble."

He shrugged, licked his lips, trembled a little. "I was lucky, I suppose. I squeezed myself into the hedge, lay still as death. And *click,* pause . . . *click,* pause, getting closer all the time. And then, behind me, for I'd turned my face away—the slow creaking as the door of the thin house swung open! And the second thin person coming out and, I imagine, unfolding himself or herself, and the two of 'em standing there for a moment, and me near dead of fright."

"And?"

"*Click-click,* pause; *click-click,* pause; *click-click*—and away they go. God only knows where they went, or what they did, but me?—I gave 'em ten minutes' start and then got up, and ran, and stumbled, and forced my rubbery legs to carry me right out of there. And I haven't been back. Why, this is the closest I've been to Barchington since that night, and too close by far!"

I waited for a moment but he seemed done. Finally I nodded. "Well, that's a good story, Bill, and—"

"I'm not finished!" he snapped. "And it's not just a story. . . ."

"There's more?"

"Evidence," he whispered. "The evidence of your own clever-bugger eyes!"

I waited.

"Go to the window," said Bill, "and peep out through the curtains. Go on, do it."

I did.

"See anything funny?"

I shook my head.

"Blind as a bat!" he snorted. "Look at the streetlights—or the absence of lights. I showed you once tonight. They've nicked all the bulbs."

"Kids." I shrugged. "Hooligans. Vandals."

"Huh!" Bill sneered. "Hooligans, here? Unheard of.

Vandals? You're joking! What's to vandalize? And when did you last see kids playing in these streets, eh?"

He was right. "But a few missing light bulbs aren't hard evidence," I said.

"All *right!*" He pushed his face close and wrinkled his nose at me. "Hard evidence, then." And he began to tell me the final part of his story. . . .

## III

"Cars!" Balmy Bill snapped, in that abrupt way of his. "They can't bear them. Can't say I blame 'em much, not on that one. I hate the noisy, dirty, clattering things myself. But tell me: have you noticed anything a bit queer—about cars, I mean—in these parts?"

I considered for a moment, replied: "Not a hell of a lot of them."

"Right!" He was pleased. "On the rest of the Hill, nose to tail. Every street overflowing. 'Specially at night when people are in the pubs or watching the telly. But here? Round Barchington and the Larches and a couple of other streets in this neighborhood? Not a one to be seen!"

"Not true," I said. "There are two cars in this very street. Right now. Look out the window and you should be able to see them."

"Bollocks!" said Bill.

"Pardon?"

"Bollocks!" he gratefully repeated. "Them's not *cars!* Rusting old bangers. Spoke wheels and all. Twenty, thirty years they've been trundling about. The thin people are *used* to them. It's the big shiny new ones they don't like. And so, if you park your car up here overnight—trouble!"

"Trouble?" But here I was deliberately playing dumb. Here I knew what he meant well enough. I'd seen it for myself: the occasional shiny car, left overnight, standing

there the next morning with its tires slashed, windows
smashed, lamps kicked in.

He could see it in my face. "You know what I mean, all
right. Listen, couple of years ago there was a flash Harry
type from the city used to come up here. There was a
barmaid he fancied in The Railway—and she was taking all
he could give her. Anyway, he was flash, you know? One of
the gang lads and a rising star. And a flash car to go with
it. Bulletproof windows, hooded lamps, reinforced panels—the lot. Like a bloody tank, it was. But—" Bill sighed.

"He used to park it up here, right?"

He nodded. "Thing was, you couldn't threaten him. You
know what I mean? Some people you can threaten, some
you shouldn't threaten, and some you mustn't. He was one
you mustn't. Trouble is, so are the thin people."

"So what happened?"

"When they slashed his tires, he lobbed bricks through
the windows. And he had a knowing way with him. He
tossed 'em through thin-house windows. Then one night he
parked down on the corner of Barchington. Next morning—they'd drilled holes right through the plate, all over
the car. After that—he didn't come back for a week or so.
When he did come back . . . well, he must've been pretty
mad."

"What did he do?"

"Threw something else—something that made a bang! A
damn big one! You've seen that thin, derelict shell on the
corner of Barchington? Oh, it was him, sure enough, and he
got it right, too. A thin house. Anybody in there, they were
goners. And *that* did it!"

"They got him?"

"They got his car! He parked up one night, went down to
The Railway, when the bar closed took his ladylove back to
her place, and in the morning—"

"They'd wrecked it—his car, I mean."

"Wrecked it? Oh, yes, they'd done that. They'd *folded* it!"

"Come again?"

"Folded it!" he snapped. "Their funny science. Eighteen inches each way, it was. A cube of folded metal. No broken glass, no split seams, no splintered plastic. Folded all neat and tidy. An eighteen-inch cube."

"They'd put it through a crusher, surely?" I was incredulous.

"Nope—folded."

"Impossible!"

"Not to them. Their funny science."

"So what did he do about it?"

"Eh? Do? He looked at it, and he thought, 'What if I'd been sitting *in* the bloody thing?' Do? He did what I would do, what you would do. He went away. We never did see him again."

The half-bottle was empty. We reached for the beers. And after a long pull I said: "You can kip here if you want, on the floor. I'll toss a blanket over you."

"Thanks," said Balmy Bill, "but no thanks. When the beer's gone I'm gone. I wouldn't stay up here to save my soul. Besides, I've a bottle of my own back home."

"Sly old sod!" I said.

"Daft young bugger!" he answered without malice. And twenty minutes later I let him out. Then I crossed to the windows and looked out at him, at the street all silver in moonlight.

He stood at the gate (where it should be) swaying a bit and waving up at me, saying his thanks and farewell. Then he started off down the street.

It was quiet out there, motionless. One of those nights when even the trees don't move. Everything frozen, despite the fact that it wasn't nearly cold. I watched Balmy Bill out of sight, craning my neck to see him go, and—

—Across the road, three lampposts—where there should

only be two! The one on the left was OK, and the one to the far right. But the one in the middle? I never had seen that one before. I blinked bleary eyes, gasped, blinked again. Only *two* lampposts!

Stinking drunk—drunk as a skunk—utterly boggled!

I laughed as I tottered from the window, switched off the light, staggered into my bedroom. The balmy old bastard had really had me going. I'd really started to believe. And now the booze was making me see double—or something. Well, just as long as it was lampposts and not pink elephants! Or thin people! And I went to bed laughing.

. . . But I wasn't laughing the next morning.

Not after they found him, old Balmy Bill of Barrows Hill. Not after they called me to identify him.

"Their funny science," he'd called it. The way they folded things. And Jesus, they'd folded him, too. Right down into an eighteen-inch cube. Ribs and bones and skin and muscles—the lot. Nothing broken, you understand, just folded. No blood or guts or anything nasty—nastier by far *because* there was nothing.

And they'd dumped him in a garbage skip at the end of the street. The couple of local youths who found him weren't even sure what they'd found, until they spotted his face on one side of the cube. But I won't go into that. . . .

Well, I moved out of there just as soon as I could—do you blame me?—since when I've done a lot of thinking about it. Fact is, I haven't thought of much else.

And I suppose old Bill was right. At least I hope so. Things he'd told me earlier, when I was only half listening. About them being the last of their sort, and Barrows Hill being the place they've chosen to sort of fade away in, like a thin person's "elephant's graveyard," you know?

Anyway, there are no thin people here, and no thin

houses. Vandals aplenty, and so many cars you can't count, but nothing out of the ordinary.

Lampposts, yes, and posts to hold up the telephone wires, of course. Lots of them. But they don't bother me anymore.

See, I know *exactly* how many lampposts there are. And I know exactly *where* they are, every last one of them. And God help the man ever plants a new one without telling me first!

# The Cyprus Shell

*The reason these next two stories are here is simple: for years
now I've wanted to use the twin stories, "The Cyprus Shell"
and "The Deep-Sea Conch," as a sort of diptych (a lovely
word, that), with the one opening a book and the other closing
it. But it never happened. And it still hasn't. So . . . why didn't
I do it here?*

*Well, it seemed only right to me that the collection's title
story should have first place, and also that "Born of the
Winds" (being the longest) should bring up the rear. Which
means that I had to squeeze these two somewhere in between.
Anyway, I did finally get them both in print together; so that
even though the shells themselves aren't bivalvular (another
lovely word), the stories finally are.*

The Oaks, Innsway,
Redcar, Yorks.
June 5, 1962

Col. (Ret'd) George L. Glee, MBE, DSO.
11 Tunstall Court,
West H. Pool, Co. Durham

My dear George,

I must extend to you my sincerest apologies for the inexcusable way I took myself and Alice out of your excellent company on Saturday evening last. Alice has remarked upon my facial *expressions,* my absolute lack of manners and the uncouth way in which, it must have appeared, I dragged her from your marvelous table; and all, alas, under the gaze of so many of our former military associates. I can only hope that our long friendship—and the fact that you know me as well as you do—has given you some insight that it was only a matter of extreme urgency that could have driven me from your house in such an extraordinary manner.

I imagine all of you were astonished at my exit. Alice was flabbergasted and would not speak to me until I gave her a solid reason for what she took to be lunatic behavior.

Well, to cut things short, I told her the tale that I am about to tell you. She was satisfied as to the validity of the reasons for my seemingly unreasonable actions and I am sure that you will feel the same.

It was the oysters, of course. I have no doubt that their preparation was immaculate and that they were delicious— for everyone, that is, except myself. The truth is I *cannot* abide seafood, especially shellfish. Surely you remember the way I used to be over crabs and lobsters? That time in Goole when I ate two whole plates of fresh mussels all to myself? I loved the things. Ugh! The thought of it . . .

Two years ago in Cyprus something happened that put an end to my appetite for that sort of thing. But before I go on let me ask you to do something. Get out your Bible and look up Leviticus 11; 10/11. No, I have not become a religious maniac. It's just that since that occurrence of two years ago I have taken a deep interest, a morbid interest I hasten to add, in this subject and all connected with it.

If, after reading my story, you should find your curiosity tickled, there are numerous books on the subject that you might like to look up—though I doubt whether you'll find many of them at your local library. Anyway, here is a list of four such books: Gantley's *Hydrophinnae,* Gaston Le Fe's *Dwellers in the Depths,* the German *Unter-Zee Kulten,* and the monstrous *Cthaat Aquadingen* by an unknown author. All contain tidbits of an almost equally nauseating nature to the tale that I must relate in order to excuse myself.

I have said it was in Cyprus. At the time I was the officer in command of a small unit in Kyrea, between Cephos and Kyrenia on the coast, overlooking the Mediterranean, that most beautiful of all oceans. In my company was a young corporal, Jobling by name, who fancied himself something of a conchologist and spent all his off-duty hours with flippers and mask snorkeling off the rocks to the south of Kyrea. I say he fancied himself, yet in fact his collection *was* quite wonderful; for he had served in most parts of the world and had looted many oceans.

Beneath glass in his billet—in beautifully made "natural" settings, all produced by his own hands—he had such varied and fascinating shells as the African *Pecten irradian,* the unicorn-horn *Murex monodon* and *Ianthina violacea* from Australia, the weird *Melongena corona* from the Gulf of Mexico, the fan-shaped *Ranella perca* of Japan, and many hundreds of others too numerous to mention here. Inevitably my weekly tour of inspection ended in Jobling's billet where I would move among his showcase marvelling at the intricacies of nature's art.

While Jobling's hobby occupied all his off-duty hours it in no way interfered with his work within the unit; he was a conscientious, hard-working N.C.O. I first noticed the *change* in him when his work began to fall off and had had it in mind for over a week to reprimand him for his slackness when he had the first of those attacks that eventually culminated so horribly.

He was found one morning, after first parade, curled on

his bed in the most curious manner, with his legs drawn up and his arms wrapped about himself—almost in a fetal position. The M.O. was called but, despite treatment, Jobling remained in his inexplicable condition for over an hour; at the end of which time he suddenly "came to" and began acting quite normally, seemingly unable to remember anything that had happened. I was obliged to relieve the man of all duties for a period of one week and he was obviously amazed at this, swearing he was fit as a fiddle and blaming his lapse on an overdose of Cyprus sun. I checked this with the M.O. who assured me that Jobling's condition had in no way corresponded to a stroke but had been, in fact, closely related to a trauma, as though the result of some deep, psychological shock. . . .

The day after he returned to duties Jobling suffered the second of his withdrawals.

This second attack took exactly the same form as the first except that it lasted somewhat longer. Also, on this occasion, he was found to be clutching in his hand a book of notes relating to his shell collection. I asked to see these notes and while Jobling, still dazed, was taken off to the hospital for an examination, I read them through in the hope of finding something that might give me some insight into the reason for his strange affliction. I had a hunch that his hobby had much to do with his condition; though just why collecting and studying shells should have such a drastic effect on anyone was beyond even guessing.

The first two-dozen pages or so were filled with observations on locations where certain species of sea snail could be found. For instance: *"Pecten irradian—small—in rock basins ten to fifteen feet deep."* Etc, etc. . . . This section was followed by a dozen pages or so of small drawings, immaculately done, and descriptions of rarer specimens. Two more pages were devoted to a map of the Kyrea coastline with shaded areas showing the locations that the collector had already explored and arrowed sections showing places still to be visited. Then, on the next page, I found a beautifully executed drawing of a shell the likes of which I had never seen before, despite my frequent studies of Jobling's showcases, and which I have seen only once since.

How to describe it? Beneath it was a scale, in inches, drawn in to show its size. It appeared to be about six inches long and its basic shape was a slender spiral: but all the way round that spiral, along the complete track from mouth to tail, were sharp spikes about two inches long at their longest and about one inch at the narrow end. They were obviously a means of defense against oceanic predators. The mouth of the thing (Jobling had colored the drawing) showed a shiny black operculum with a row of tiny eyes at the edges, like those of the scallop, and was a bright shade of pink. The main body of the shell and the spikes were sand colored. If my powers of description were better I might be able to convey something of just how repulsive the thing looked. Instead it must suffice to say that it was not a shell that I would be happy to pick off a beach, and not just because of those spikes! There was something nastily fascinating about the *shape* and the *eyes* of the thing, which, taking into consideration the accuracy of the other drawings, was not merely a quirk of the artist. . . .

The next page was a set of notes, which, as best I can remember, went like this:

*Murex hypnotica?*
*Rare? Unknown???*
*August 2nd* . . . Found shell, with snail intact, off point of rocks (marked on sketch map) in about twenty feet of water. The shell was on sand in natural rock basin. Pretty sure thing is very rare, probably new species. Snail has eyes on edge of mantle. Did not take shell. Anchored it to rock

with nylon line from spear gun. Want to study creature in natural surroundings before taking it.

*August 3rd* . . . Shell still anchored. Saw most peculiar thing. Small fish, inch long, swam up to shell, probably attracted by bright pink of mouth. Eyes of snail waved rhythmically for a few seconds. Operculum opened. *Fish swam into shell and operculum snapped shut.* Have named shell well. First funny fascination I felt when I found shell is obviously felt to greater extent by fish. Thing seems to use hypnosis on fish in same way octopus uses it to trap crabs.

*August 4th* . . . Cannot visit shell today—duty. Had funny dream last night. I was in a shell on bottom of sea. Saw *myself* swimming down. Hated the swimmer and saw him as being to blame for restricting my movements. *I was the snail!* When the swimmer, myself, was gone I sawed at nylon line with my operculum but could not break it. Woke up. Unpleasant.

*August 5th* . . . Duty.

*August 6th* . . . Visited shell again. Line near shell slightly frayed. Eyes waved at me. Felt dizzy. Stayed down too long. Came back to camp. Dizzy all day.

*August 7th* . . . Dreamed of being snail again. Hated swimmer, myself, and tried to get *into* his mind—like I do with fish. Woke up. Awful.

There was a large gap in dates here and, looking back, I realized that this was the period that Jobling had spent in hospital. In fact he had gone into hospital the day after that last entry had been made, on the eighth of the month. The next entry went something like this:

*August 15th* . . . Did not expect shell to be there after all this time but it was. Line much frayed but not broken. Went down five or six times but started to get horribly dizzy. Snail writhed its eyes at me frantically! *Felt that awful dream*

*coming on in water!* Had to get out of sea. Believe the damn
thing tried to hypnotize me like it does with fish.
   *August 16th* . . . Horrible dream. Have just woke up and
must write it down now. *I was the snail again!* I've had
enough. Will collect shell today. This dizzy feeling . . .

   That was all. The last entry had obviously been made
that very day, just before Jobling's new attack. I had just
finished reading the notes and was sitting there bewildered
when my telephone rang. It was the M.O. All hell was on
at the hospital. Jobling had tried to break loose, tried to get
out of the hospital. I took my car straight round there and
that was where the horror really started.
   I was met by an orderly at the main entrance and es-
corted up to one of the wards. The M.O. and three male
nurses were in a wardroom waiting for me.
   Jobling was curled up on a bed in that weird fetal position—*or was it a fetal position?* What did he remind me of?
Suddenly I noticed something that broke my train of
thought, causing me to gasp and look closer. Froth was
drying on the man's mouth, his teeth were bared, and his
eyes bulged horribly. But there was no movement in him at
all! *He was stone dead!*
   "How on Earth? . . ." I gasped. "What happened?" The
M.O. gripped my arm. His eyes were wide and unbelieving.
For the first time I noticed that one of the male nurses
appeared to be in a state of shock. In a dry, cracked voice
the M.O. started to speak.
   "It was horrible. I've never known anything like it. He
just seemed to go wild. Began frothing at the mouth and
tried to get out of the place. He made it down to the main
doors before we caught up to him. We had to *carry* him up
the stairs and he kept straining toward the windows in the
direction of the sea. When we got him back up here he
suddenly coiled up—just like that!" He pointed to the still

figure on the bed. "And then he squirmed—that's the only way I can describe how he moved—he *squirmed* off the bed and quick as a flash he was in the steel locker there and had pulled the door shut behind him. God only knows why he went in there!" In one corner of the room stood the locker. One of its two doors was hanging by a wrenched hinge— torn almost completely off. "When we tried to get him out he fought like hell. But not like a *man* would fight! He *butted* with his head, with a funny *sawing* motion, and bit and spat—and all the time he stayed in that awful position, even when he was fighting. By the time we got him to the bed again he was dead. I . . . I think he died of fright."

By now that hideous train of thought, broken before by the horror of Jobling's condition, had started up once more in my mind and I began to trace an impossible chain of events. But no! It was too monstrous to even think of—too *fantastic* . . . And then, on top of my terrible thoughts, came those words from the mouth of one of the nurses that caused me to pitch over suddenly into the darkness of a swoon.

I know you will find it difficult to believe, George, when I tell you what those words were. It seems ridiculous that such a simple statement could have so deep an effect upon anyone. Nontheless I *did* faint; for I saw the sudden connection, the piece in the puzzle that brought the whole picture into clear, horrific perspective. . . . The nurse said:

"Getting him out of the locker was the worst, sir. *It was like trying to get a winkle out of its shell without a pin. . . .*"

When I came to, despite the M.O.'s warning, I went to the mess—I had not met Alice at that time and so was "living in"—and got my own swimming kit out of my room. I took Jobling's notebook with me and drove to the point of rock marked in red on his map. It was not difficult to find the place; Jobling would have made an excellent cartographer.

I parked my car up and donned my mask and flippers. In no time at all I was swimming straight out in the shallows over a few jagged outcrops of rock and patches of sand. I stopped in about ten feet of water for a few seconds to watch a heap, literally a *heap,* of crabs fighting over a dead fish. The carcass was completely covered by the vicious things—it's amazing how they are attracted by carrion—but occasionally, as the fight raged, I caught a glimpse of silver scales and red, torn flesh. But I was not there to study the feeding habits of crabs. I pressed on.

I found the rock basin almost immediately and the shell was not difficult to locate. It was lying about twenty feet deep. I could see that the nylon line was still attached. But for some reason the water down there was not as clear as it should have been. I felt a sudden, icy foreboding—a nameless premonition. Still, I had come out to look at that shell, to prove to myself that what I had conjured up in my overimaginative mind was pure fancy and nothing more.

I turned on end, pointing my feet at the blue sky and my head toward the bottom, and slid soundlessly beneath the surface. I spiraled down to the shell, noting that it was exactly as Jobling had drawn it, and carefully, shudderingly, took it by one of those spikes and turned it over so I could see the pink mouth.

The shell was *empty!*

But this, if my crazy theory was correct, was just what I should have expected; nonetheless I jerked away from the thing as though it had suddenly become a conger eel. Then, out of the corner of my eye, I saw the *reason* for the murkiness of the water. A second heap of crabs was sending up small clouds of sand and sepia from some dead fish or mollusk that they were tearing at in that dreadful, frantic lust of theirs.

Sepia! In abrupt horror I recoiled from my own thoughts. Great God in heaven! *Sepia!*

I wrenched off a flipper and batted the horrid, scrabbling things away from their prey—*and wished immediately that I had not done so.* For sepia is the blood, or juice, of cuttle-fish *and certain species of mollusk and sea snail.*

The thing was still alive. Its mantle waved feebly and those eyes that remained saw me. Even as the ultimate horror occurred I remember I suddenly knew for certain that my guess had been correct. For the thing was not *coiled* like a sea snail should be—*and what sea snail would ever leave its shell?*

I have said it saw me! George, I swear on the Holy Bible that the creature *recognized* me and, as the crabs surged forward again to the unholy feast, it tried to *walk* toward me.

Snails should not walk, George, and men should not squirm.

Hypnotism is a funny thing. We barely understand the *human* form of the force let alone the strange strains used by lesser life-forms.

What more can I say? Let me just repeat that I hope you accept my apologies for my behavior the other night. It was the oysters, of course. Not that I have any doubt that their preparation was immaculate or that they were delicious—to any *untainted* palate, that is. But myself? Why! I could no more eat an oyster than I could a corporal.

Sincerely,
Maj. Harry Winslow

# The Deep-Sea Conch

*Ramsey Campbell called "Conch" "nasty." What more recommendation do you need?*

11 Tunstall Court,
West H. Pool,
Co. Durham
June 16th 1962

Maj. Harry Winslow,
The Oaks, Innsway,
Redcar, Yorks.

My dear Harry,

What an interesting tale that letter of yours tells. Hypnotic gastropods, by crikey! But there was no need for the apologies, honestly. Why! We all fancied you were ill or something, which you were as it happened, and so we didn't for an instant consider your actions as being "inexcusable." If only I'd known you were so, well, *susceptible* to shellfish; I would never have had oysters on the card! And for goodness sake, do put Alice's mind at rest and tell her that there's nothing to forgive. Give her my love, poor darling!

But that story of yours is really something. I showed your letter to a conchologist friend of mine from Harden. This fellow—John Beale's his name—spends all his summers down south, "conching," as he calls it, in the coves of Devon and Dorset; or at least he *used* to. Why! He was as keen a chap as your Corporal Jobling (poor fellow) seems to have been.

But it appears I must have a stronger stomach than you, Old Chap. Yes, indeed! In fact I've just been down to The Lobster Pot on the seafront for a bite and a bottle, and I couldn't resist a fat crab straight out of the sea at Old Hartlepool. Yes, and that after both your story *and* this tale of Beale's. Truth is, I can't make up my mind which of the two stories is the more repulsive—but I might add that neither of them appears to have affected my appetite!

Still, I've just got to tell you this other story—just as Beale told it to me—so, taking a chance on further offending your sensibilities, here we go:

Now you'll probably remember how a few years ago a British oceanographical expedition charted the continental shelf all the way up from the Bay of Biscay along the North Atlantic Drift to the Shetland Isles. I say you'll probably remember, because of course the boat went down off the Faroes right at the end of its voyage, and it was in all the newspapers at the time. Luckily, lifeboats got the crew of *The Sunderland* off before she sank.

Well anyway, one of the crew was a friend of this Beale chap and he knew he was keen on shells and so on. He brought back with him a most peculiar specimen as a present for Beale. It was a deep-sea conch, dredged up from two thousand seven hundred fathoms beneath the Atlantic two hundred miles west of Brest.

Now this thing was quite a find, and Beale's friend would have found himself in plenty of hot water had the professors aboard ship found out how he'd taken it from the dredge. As it was he kept it hidden under his bunk until the boat went down, and even then he managed to smuggle it into a lifeboat with him. And so he brought it home. Of course, after a month out of the water—even had it survived the emergence from such a depth—the creature in the conch would be long dead.

You'd think so, wouldn't you, Harry? . . . As soon as Beale got hold of the conch he stuck it in a bowl of weak acid solution to clean up its surface and get rid of the remains of the animal inside. It was quite large, ten or eleven inches across and four inches deep, tightly coiled, and with a great bell of a mouth. Beale reckoned that once he'd got the odd oceanic incrustations off the thing it ought to be rather exotically patterned. Most deep-ocean dwellers are exotic, in their way, you know?

The next morning, Beale went to take the conch out of the acid—and as he did so he noticed that a great, thick, shiny-green operculum was showing deep down in the bell-shaped mouth. The acid hadn't even touched the incrustations on the shell, but it had obviously loosened up the dead snail (is that what you call them?) inside the tight coil. Which was where the fun started.

Beale took a knife and tried to force the blade down between the green lid and the interior of the hard shell, so that he could hook out the body of the snail, d'you see? But as soon as the knife touched the operculum—*damn it if the thing didn't withdraw!* Yes, it was still alive, that creature— even after being brought up from that tremendous depth, even after a month out of water and a night in that bowl of acid solution—still alive! Fantastic, eh, Harry?

Now, Beale knew he had a real curiosity here, and that he should take it at once to some zoologist or other. But how could he without dropping his friend right in it? Why! The deep-sea conch looked like being the best thing to come out of the whole abortive expedition. . . .

Well, Beale thought about it, and a week later he finally decided to get a second opinion. He asked his oceangoing friend (a Hartlepool man, by the way, name of Chadwick) up to Harden where he brought him up to date before asking his advice. By then he'd bought an aquarium for the conch and was feeding it on bits of raw meat. Very unpleasant. He never saw it actually *take* the meat, you understand, but each morning the tank would be empty except for the great shell and its occupant.

Beale's chief hope in asking Chadwick for his advice was that the fellow'd tell him to hand the conch in. He hoped, you see, that Chadwick would be willing to face the music in "the interests of science." Not so. This Chadwick was simply an ordinary little Jack-Tar, and he feared the comeback of owning up to his bit of thievery. Why couldn't Beale (he wanted to know) simply keep the thing and stop worrying about it? Well, Beale explained that his hobby was conchology, not zoology, and that he didn't fancy having this aquarium-thing smelling up his flat forever, and so on. At which Chadwick argued why couldn't he just kill off the snail and keep the shell? And Beale had no answer to that.

He's not a bad chap, this Beale, you see, and he knew very well what he really ought to do—that is, he should hand the conch over and let Chadwick take his medicine— but damn it all, the fellow was his friend! In the end he agreed to kill the snail and keep the shell. That way he could get rid of the smelly aquarium—the conch did have quite a bit of a pong about it—and live happily ever after.

But Chadwick had seen Beale's initial indecision over the conch's destiny, and he decided he'd better stick around and see to it that his pal really did kill the snail off—and that's where things started to get a bit complicated. That night, by the time Chadwick was ready to get off back to Hartlepool, they still hadn't managed to put the damn thing out of its misery. And believe me, Harry, they'd really tried!

First off Beale had taken the conch from the aquarium to place it in a powerful acid bath. Two hours later the deposits on the shell had vanished, allowing a disappointing pattern to show through. Also, the surface of the acid had scummed over somewhat—but the hard, shiny-green operculum was still there, tight as the doors on the vaults at the Bank of England! Now remember, our friends knew that the snail could get along fine for at least a month without food or water, and it looked as though the lid and the hard shell were capable of withstanding the strongest acid that Beale could get hold of. So what next?

Even after his friend had gone off home Beale was still puzzling over the problem. A damned weird thing this altogether, and he was starting to feel sort of uneasy about it. It reminded him of something, this shell, something he'd seen somewhere before . . . at the local museum, for instance!

The next morning, as soon as his sailor friend turned up, they went off together down to the museum. In the "Prehistoric Britain" section Beale found what he was looking for—a whole shelf of fossil shells of all sorts, shapes, and sizes. Of course, the fossils were all under glass so that our pair couldn't handle them, but they didn't really need to. "There you go," said Beale, excitedly pointing out one of the specimens, "that's it—not so big, perhaps, but the same shape and with the same type of markings!" And Chadwick had to agree that the fossil looked very much similar to their conch.

Well, they read off the card behind the fossil, and here, (as best I can remember from what Beale told me—though I'm not absolutely sure about the name, as you can see,) is what it said:

Upper Cretaceous:
. (——?——SCAPHITES, a tightly coiled shell from Bar-

row-on-Soar in Leicestershire. Flourished in Cretaceous oceans 120 million years ago. Similar specimens, with more prominent ribs and nodes, occur frequently in the Leicestershire Lias. Extinct for over 60 million years . . .

. . . Which didn't tell them very much, you'll agree, Harry. But Beale jumped a bit at that "extinct" notice—he had reckoned it might be so—and once again he asked Chadwick to turn the thing in. Why! It was starting to look as good as the coelacanth!

"No hope!" Chadwick answered, perfunctorily. "Let's get round to your place and finish the thing off. For the last time, I'm not carrying the can for doing you a favor!" And so they went back to Beale's flat.

There, waiting for them in the kitchen, was the conch—still in its bath of acid, and apparently completely unharmed. Well, they carefully drained off the acid and sprayed the conch down with water to facilitate its immediate handling. Chadwick had brought with him from Hartlepool a great knife with a hooked blade. He tried his best to get the point right down inside the bell of the shell, but always there was that incredibly hard operculum, all shiny green and tight as a cork in a bottle.

This was when Chadwick, having completely lost patience with it all, suggested smashing the conch. Suggested it? Why! He had the thing out on the concrete landing before Beale could get his thoughts in order. But the conchologist needn't have worried; I mean, how does one go about smashing something that's survived the pressure at two thousand seven hundred fathoms, eh? Chadwick had actually flung the conch full swing against the concrete floor of the landing by the time Beale caught up with him, and the latter was just able to catch it on the bounce. It wasn't even dented! Peering into the bell-shaped mouth, Beale was just able to make out one edge of the green operculum pulling back out of sight around the curve of the coil.

Then Chadwick had his brain wave. One thing the conch had *not* been subjected to down there on the bed of the ocean was heat—it's a monstrous cold world at the bottom of the sea! And it just so happened that Beale had already made inquiries at a local hardware store about borrowing a blowtorch to lift the paint from a wardrobe he wanted to do up. Chadwick went off to borrow the blowtorch and left Beale sitting deep in thought contemplating the conch.

Now, I've already mentioned how Beale had had this strange, uneasy feeling about the shell and its occupant. Yes; well by now the feeling had grown out of all proportion. There seemed to the sensitive collector to be a sort of, well, an *aura* about the thing, that feeling of untold ages one has when gazing upon ancient ruins—except that with the conch the sensation was far more powerful.

And then again, how had the conch come by its amazing powers of self-preservation? Was it possible? . . . No, what Beale was thinking was plainly impossible—such survivals were out of the question—and yet the mad thought kept spinning around in Beale's brain that perhaps, perhaps . . .

How *long* had this creature, encased in the coils of its own construction (Beale's words), prowled the pressured deeps in ponderous stealth? "Extinct for 60 million years . . ." Suddenly he dearly wished he had some means at his disposal of checking the conch's exact age. It was a crazy thought, he knew, but there was this insistent idea in the back of his head—

In his mind's eye he saw the world as it had been so long ago—great beasts trampling primitive plants in steaming swamps, and strange birds that were *not* birds flying in heavy, predawn skies. And then he looked beneath those prehistoric oceans—oceans more like vast *acidbaths* than seas such as we know them today—at the multitude of forms that swam, spurted, and crawled in those deadly deeps.

Then Beale allowed (as he put it) an "unwinding of time" in his mind, picturing the geologic changes, seeing continents emerge hissing from volcanic seas, and coralline islands slowly sinking into the hoary soups of their own genesis. He watched the gradual alteration in climates and environments, and the effects such changes had on their denizens. He saw the remote forbears of the conch altering internally to build a resistance to the tremendous pressures of even deeper seas, and, as their numbers dwindled, developing fantastic life spans to ensure the continuation of the species.

Beale told me all this, you understand, Harry? And God only knows where his thoughts might have led him if Chadwick hadn't come back pretty soon with the blowtorch. But even after the sailor got the blowtorch going—while he played its terrible tip of invisible heat on the coils of the conch—Beale's fancy was still at work.

Once, as a boy, he had pulled a tiny hermit crab from its borrowed shell on the beach at Seaton-Carew, to watch it scurry in frantic terror in search of a new home over the sand at the bottom of a small pool. In the end, out of an intensely agonizing empathy for the completely vulnerable crab, he had dropped the empty whelk shell back into the water in the soft-bodied creature's way, so that it was able to leap with breathtaking rapidity and almost visible relief back into the safety of the shell's calcium coil. It's death for this type of creature to be forced from its protective shell, you see, Harry? That's what Beale was telling me, and once out such a creature has only two alternatives—get back in or find a new home . . . and quick, before the predators come on the scene! Small wonder the snail in the conch was giving them such a hard time!

And that was when Beale heard Chadwick's hiss of indrawn breath and his shouted, "It's coming out!"

Beale had been turned away from what he rightly considered a criminal scene, but at Chadwick's cry he turned back—in time to see the sailor drop the blowtorch and fling himself almost convulsively away from the metal sink unit where he'd been working. Chadwick said nothing more, you understand, following that initial cry. He simply hurled himself away from the steaming conch.

Quick as thought, the flame from the fallen blowtorch caught at the kitchen tablecloth, and Beale's first impulse was to save his flat. Chadwick had burned himself, that was plain, but his burn couldn't be all that serious. So thinking, Beale leapt over to the sink to fill a jug with water. The torch was still flaring but it lay on the ceramic-tiled floor where it could do little harm. As he filled his jug Beale noticed how the overspill moved the conch, as though the shell were somehow lighter, but he had no time to ponder that.

Turning, he almost dropped the jug as he saw Chadwick's figure stretched full length half-in, half-out of the kitchen. Dodging the leaping flames and futilely flinging the jug's contents over them, he crossed to where his friend lay, turning him onto his back to see what was wrong. There were no signs of any burns, but, quickly checking, Beale was horrified at his inability to detect a pulse.

Shock! Chadwick must be suffering from shock! Beale pushed his fingers into the sailor's mouth to loosen up his tongue, and then, seeing a slight movement of the man's throat, he threw himself face down beside him to get into the "Kiss of Life" position.

He never administered that kiss, but leapt shrieking to flee from the burning flat, down the stairs and out of the building. He told me that apart from that mad dash he can remember nothing more of the nightmare—nothing, that is, other than the *cause* of his panic flight. Of course, there's no proving Beale's story one way or the other; the whole block of flats was burned right out and it was a miracle that Chadwick was the only casualty. I don't suppose I'd ever have got the story out of Beale if I hadn't met up with him one night and showed him that letter of yours. He was almighty drunk at the time, and I'm sure he's never told anyone else.

Fact or fiction? . . . Damned if I know, but it's true that Beale doesn't go "conching" any more, and just suppose it *did* happen the way he told it?

What a shock to a delicate nervous system, eh, Harry?

You can probably guess what happened, but just put yourself in Beale's place. Imagine the heart-stopping horror when, having seen Chadwick's throat move, *he looked into the sailor's mouth and saw that shiny-green operculum pull down quickly out of sight into the fellow's throat!*

# Born of the Winds

*Way back in 1976, this final inclusion was nominated for a World Fantasy Award. It didn't win, but I wasn't unhappy— because I didn't know! Nobody had bothered to tell me. In fact I didn't find out until ten years later in Providence, Rhode Island, where I was given a copy of the 1986 World Fantasy Convention program book, containing "An Outline History of the World Fantasy Convention." That's where I saw it for the first time in black and white. And what do you know, I still wasn't unhappy. Because the winner back in 1976 had been Fritz Leiber's "Belsen Express." To have something mentioned in the same breath means you're doing OK.*

*Of my many Lovecraft- or Mythos-oriented stories, this one is probably my favorite.*

## I

Consider: I am, or was, a meteorologist of some note—a man whose interests and leanings have always been away from fantasy and the so-called "supernatural"— and yet now I believe in a wind that blows between the worlds, and in a Being that inhabits that wind, striding in feathery cirrus and shrieking lightning storm alike across icy Arctic heavens.

Just how such an utter *contradiction* of beliefs could come
about I will now attempt to explain, for I alone possess all
of the facts. If I am wrong in what I more than suspect—if
what has gone before has been nothing but a monstrous
chain of coincidence confused by horrific hallucination—
then with luck I might yet return out of this white wilderness to the sanity of the world I knew. But if I am right, and
I fear that I am horribly right, then I am done for, and this
manuscript will stand as my testimonial of a hitherto all but
unrecognized plane of existence . . . and of its *inhabitant,*
whose like may only be found in legends whose sources date
back geological eons into Earth's dim and terrible infancy.

My involvement with this thing has come about all in the
space of a few months, for it was just over two months ago,
fairly early in August, that I first came to Navissa, Manitoba, on what was to have been a holiday of convalescence
following a debilitating chest complaint.

Since meteorology serves me both as hobby and means of
support, naturally I brought some of my "work" with me;
not physically, for my books and instruments are many, but
locked in my head were a score of little problems beloved
of the meteorologist. I brought certain of my notebooks,
too, in which to make jottings or scribble observations on
the almost Arctic conditions of the region as the mood
might take me. Canada offers a wealth of interest to one
whose life revolves about the weather: the wind and rain,
the clouds, and the storms that seem to spring from them.

In Manitoba on a clear night, not only is the air sweet,
fresh, sharp, and conducive to the strengthening of weakened lungs, but the stars stare down in such crystal clarity
that at times a man might try to pluck them out of the
firmament. It is just such a night now—though the glass is
far down, and I fear that soon it may snow—but warm as
I am in myself before my stove, still my fingers feel the

awesome cold of the night outside, for I have removed my
gloves to write.

Navissa, until fairly recently, was nothing more than a trail
camp, one of many to expand out of humble beginnings as
a trading post into a full-blown town. Lying not far off the
old Olassie Trail, Navissa is quite close to deserted, ill-fated
Stillwater; but more of Stillwater later. . . .

I stayed at the judge's house, a handsome brick affair
with a raised log porch and chalet-style roof, one of
Navissa's few truly modern buildings, standing on that side
of the town toward the neighboring hills. Judge Andrews is
a retired New Yorker of independent means, an old friend
of my father, a widower whose habits in the later years of
his life have inclined toward the reclusive; being self-suffi-
cient, he bothers no one, and in turn he is left to his own
devices. Something of a professional anthropologist all his
life, the judge now studies the more obscure aspects of that
science here in the thinly populated North. It was Judge
Andrews himself, on learning of my recent illness, who so
kindly invited me to spend this period of convalescence with
him in Navissa, though by then I was already well on the
way to recovery.

Not that his invitation gave me license to intrude upon
the judge's privacy. It did not. I would do with myself what
I would, keeping out of his way as much as possible. Of
course, no such arrangement was specified, but I was aware
that this was the way the judge would want it.

I had free run of the house, including the old gentleman's
library, and it was there one afternoon early in the final
fortnight of my stay that I found the several works of
Samuel R. Bridgeman, an English professor of anthropology whose mysterious death had occurred only a few dozen
miles or so north of Navissa.

Normally such a discovery would have meant little to me,

but I had heard that certain of Bridgeman's theories had made him something of an outcast among others of his profession; there had been among his beliefs some that belonged in no way to the scientific. Knowing Judge Andrews to be a man who liked his facts straight on the line, undistorted by whim or fancy, I wondered what there could be in the eccentric Bridgeman's works that prompted him to display them upon his shelves.

In order to ask him this very question, I was on my way from the small library room to Judge Andrews's study when I saw, letting herself out of the house, a distinguished-looking though patently nervous woman whose age seemed rather difficult to gauge. Despite the trimness of her figure and the comparative youthfulness of her skin, her hair was quite grey. She had plainly been very attractive, perhaps even beautiful, in youth. She did not see me, or if she did glimpse me where I stood, then her agitated condition did not admit of it. I heard her car pull away.

In the doorway of the judge's study I formed my question concerning Bridgeman's books.

"Bridgeman?" the old man repeated me, glancing up sharply from where he sat at his desk.

"Just those books of his, in the library," I answered, entering the room proper. "I shouldn't have thought that there'd be much for you, judge, in Bridgeman's work."

"Oh? I didn't know you were interested in anthropology, David?"

"Well, no, I'm not really. It's just that I remember hearing a thing or two about this Bridgeman, that's all."

"Are you sure that's all?"

"Eh? Why, certainly! Should there be more?"

"Hmm," he mused. "No, nothing much—coincidence. You see, the lady who left a few moments ago was Lucille Bridgeman, Sam's widow. She's staying at the Nelson."

"Sam?" I was immediately interested. "You knew him then?"

"I did, fairly intimately, though that was many years ago. More recently I've read his books. Did you know that he died quite close by here?"

I nodded. "Yes, in peculiar circumstances, I gather?"

"That's so, yes." He frowned again, moving in his chair in what I took to be agitation.

I waited for a moment, and then when it appeared that the judge intended to say no more, I asked, "And now?"

"Hmm?" His eyes were far away even though they looked at me. They quickly focused. "Now—nothing . . . and I'm rather busy!" He put on his spectacles and turned his attention to a book.

I grinned ruefully, inclined my head, and nodded. Being fairly intimate with the old man's moods, I knew what his taciturn, rather abrupt dismissal had meant: "If you want to know more, then you must find out for yourself!" And what better way to discover more of this little mystery, at least initially, than to read Samuel R. Bridgeman's books? That way I should at least learn something of the man.

As I turned away, the judge called to me: "Oh, and David—I don't know what preconceptions you may have formed of Sam Bridgeman and his work, but as for myself . . . near the end of a lifetime, I'm no closer now than I was fifty years ago to being able to say what *is* and what *isn't*. At least Sam had the courage of his convictions!"

What was I to make of that?—and how to answer it? I simply nodded and went out of the room, leaving the judge alone with his books and his thoughts. . . .

That same afternoon found me again in the library, with a volume of Bridgeman's on my lap. There were three of his books in all, and I had discovered that they contained many references to Arctic and near-Arctic regions, to their

people, their gods, superstitions, and legends. Still ponder-
ing what little I knew of the English professor, these were
the passages that primarily drew my attention: Bridgeman
had written of these northern parts, and he had died here—
mysteriously! No less mysterious, his widow was here now,
twenty years after his demise, in a highly nervous if not
actually distraught state. Moreover, that kindly old family
friend Judge Andrews seemed singularly reticent with regard to the English anthropologist, and apparently the
judge did not entirely disagree with Bridgeman's controversial theories.

But what were those theories? If my memory served me
well, then they had to do with certain Indian and Eskimo
legends concerning a god of the Arctic winds.

At first glance there seemed to be little in the professor's
books to show more than a normally lively and entertaining
anthropological and ethnic interest in such legends, though
the author seemed to dwell at unnecessary length on Gaoh
and Hotoru, air-elementals of the Iroquois and Pawnee
respectively, and particularly upon Negafok, the Eskimo
cold-weather spirit. I could see that he was trying to tie such
myths in with the little-known legend of the Wendigo, of
which he seemed to deal far too positively.

"The Wendigo," Bridgeman wrote, "is the avatar of a
Power come down the ages from forgotten gulfs of immemorial lore; this great *Tornasuk* is none other than
Ithaqua Himself, the Wind-Walker, and the very sight of
Him means a freezing and inescapable death for the unfortunate observer. Lord Ithaqua, perhaps the very greatest of
the mythical air-elementals, made war against the Elder
Gods in the Beginning; for which ultimate treason He was
banished to frozen Arctic and interplanetary heavens
to 'Walk the Winds Forever' through fantastic cycles of
time and to fill the *Esquimaux* with dread, eventually ear-
ning His terrified worship and His sacrifices. None but such

worshipers may look upon Ithaqua—for others to see Him is certain death! He is as a dark outline against the sky, anthropomorphic, a manlike yet bestial silhouette, striding both in low icy mists and high stratocumulus, gazing down upon the affairs of men with carmine stars for eyes!" Bridgeman's treatment of the more conventional mythological figures was less romantic; he remained solidly within the framework of accepted anthropology. For example: "The Babylonian storm-god, Enlil, was designated 'Lord of the Winds.' Mischievous and mercurial in temperament, he was seen by the superstitious peoples of the land to walk in hurricanes and sand-devils. . . ." Or, in yet more traditional legend: "Teuton mythology shows Thor as being the god of thunder; when thunderstorms boiled and the heavens roared, people knew that what they heard was the sound of Thor's war-chariot clattering through the vaults of heaven."

Again, I could not help but find it noticeable that while the author here poked a sort of fun at these classical figures of mythology, he had *not* done so when he wrote of Ithaqua. Similarly, he was completely dry and matter-of-fact in his descriptive treatment of an illustration portraying the Hittite god-of-the-storm, Tha-thka, photographed from his carved representation upon a baked clay tablet excavated in the Toros Mountains of Turkey. More, he compared Tha-thka with Ithaqua of the Snows, declaring that he found parallels in the two deities other than the merely phonetical similarity of their names.

Ithaqua, he pointed out, had left webbed tracks in the Arctic snows, tracks that the old *Esquimaux* tribes feared to cross; and Tha-thka (carved in a fashion very similar to the so-called "Amarna style" of Egypt, to mix ethnic art groups) was shown in the photograph as having star-shaped eyes of a rare, dark carnelian . . . and webbed feet! Professor Bridgeman's argument for connection here seemed valid, even sound, yet I could see how such an argument might very well anger established anthropologists of "the Old School." How, for instance, might one equate a god of the ancient Hittites with a deity of comparatively modern Eskimos? Unless of course one was to remember that in a certain rather fanciful mythology Ithaqua had only been banished to the North following an abortive rebellion against the Elder Gods. Could it be that *before* that rebellion the Wind-Walker strode the high currents and tides of atmospheric air over Ur of the Chaldees and ancient Khem, perhaps even prior to those lands being named by their first inhabitants? Here I laughed at my own fancies, conjured by what the writer had written with such assumed authority, and yet my laughter was more than a trifle strained, for I found a certain cold logic in Bridgeman that made even his wildest statement seem merely a calm, studied exposition. . . .

And there were, certainly, wild statements. The slimmest of the three books was full of them, and I knew after reading only its first few pages that this must be the source of those flights of fancy that had caused Bridgeman's erstwhile colleagues to desert him. Yet without a doubt the book was by far the most interesting of the three, written almost in a fervor of mystical allusion with an abundance—a *plethora*—of obscure hints suggestive of half-discernible worlds of awe, wonder, and horror bordering and occasionally impinging upon our very own.

I found myself completely enthralled. It seemed plain to me that behind all the hocus-pocus there was a great mystery here—one that, like an iceberg, showed only its tip—and I determined not to be satisfied with anything less than a complete verification of the facts concerning what I had started to think of as "the Bridgeman case." After all, I seemed to be ideally situated to conduct such an investigation: this was where the professor had died, the borderland of that region in which he had alleged at least one of his mythological beings to exist; and Judge Andrews, provided I could get him to talk, must be something of an authority on the man; and, possibly my best line of research yet, Bridgeman's widow herself was here now in this very town.

Just why this determination to dabble should have so enthused me I still cannot say; unless it was the way that Tha-thka, which Being Bridgeman had equated with Ithaqua, was shown upon the Toros Mountains tablet as walking splayfooted through a curious mixture of cumulonimbus and nimbostratus—cloud formations that invariably presage snow and violent thunderstorms! The ancient sculptor of that tablet had certainly gauged the Wind-Walker's domain well, giving the mythical creature something of solidarity in my mind, though it was still far easier for me to accept those peculiar clouds of ill omen than the Being striding among them. . . .

## II

It was something of a shock for me to discover, when finally I thought to look at my wristwatch, that Bridgeman's books had kept me busy all through the afternoon and it was now well into evening. I found that my eyes had started to ache with the strain of reading as it grew darker in the small library room. I put on the light and would have returned to the books yet again but for hearing, at the outer door of the house, a gentle knocking. The library door was slightly ajar so that I could hear the judge answering the knocking and his gruff welcome. I was sure that the voice that answered him was that of Bridgeman's widow, for it was vibrant with a nervous agitation as the visitor entered the house and went with the judge to his study. Well, I had desired to meet her; this seemed the perfect opportunity to introduce myself.

Yet at the open door to the judge's study I paused, then quickly stepped back out of sight. It seemed that my host and his visitor were engaged in some sort of argument. He had just answered to some unheard question: "Not *me,* my dear, that is out of the question. . . . But if you insist upon this folly, then I'm sure I can find someone to help you. God knows I'd come with you myself—even on this wild-goose chase you propose, and despite the forecast of heavy snow—but . . . my dear, I'm an old man. My eyes are no good anymore; my limbs are no longer as strong as they used to be. I'm afraid that this old body might let you down at the worst possible time. It's bad country north of here when the snows come."

"Is it simply that, Jason," she answered him in her nervous voice, "or is it really that you believe I'm a madwoman? That's what you as good as called me when I was here earlier."

"You must forgive me for that, Lucille, but let's face it—that story you tell is simply . . . *fantastic!* There's no positive proof that the boy headed this way at all, just this premonition of yours."

"The story I told you was the truth, Jason! As for my 'premonition,' well, I've brought you proof! Look at this—"

There was a pause before the judge spoke again. Quietly he asked, "But what is this thing, Lucille? Let me get my glass. Hmm—I can see that it depicts—"

*"No!"* Her cry, shrill and loud, cut him off. "No, don't mention *Them,* and please don't say His name!" The hysterical emphasis she placed on certain words was obvious, but she sounded calmer when, a few seconds later, she continued: "As for what it is—" I heard a metallic clinking, like a coin dropped on the tabletop, "just keep it here in the house. You will see for yourself. It was discovered clenched in Sam's right hand when they—when they found his poor, broken body."

"All that was twenty years ago—" the judge said, then paused again before asking: "Is it gold?"

"Yes, but of unknown manufacture. I've shown it to three or four experts over the years, and always the same answer. It is a very ancient thing, but from no known or recognizable culture. Only the fact that it is made of gold saves it from being completely alien! And even the gold is . . . not quite right. Kirby has one, too."

"Oh?" I could hear the surprise in the judge's voice. "And where did he get it? Why, just looking at this thing under the glass, I should have taken it for granted—even knowing nothing of it—that it's as rare as it's old!"

"I believe they are very rare indeed, surviving from an age before all earthly ages. Feel how cold it is. It has a chill like the ocean floor, and if you try to warm it . . . but try it for yourself. I can tell you now, though, that it will not *stay* warm. And I know what that means. . . .

"Kirby received his in the mail some months ago, in the summer. We were at home in Mérida, in Yucatán. As you know, I settled there after—after—"

"Yes, yes I know. But who would want to send the boy such a thing—and why?"

"I believe it was meant as—as a *reminder,* that's all—as a means to awaken in him all I have worked to keep dormant. I've already told you about . . . about Kirby, about his strange ways even as a baby. I thought they would leave him as he grew older. I was wrong. That last month before he vanished was the worst. It was after he received the talisman through the mail. Then, three weeks ago, he—he just packed a few things and—" She paused for a moment, I believed to compose herself, for an emotional catch had developed in her voice. I felt strangely moved.

"—As to who sent it to him, that's something I can't say. I can only guess, but the package carried the Navissa postmark! That's why I'm here."

"The Navissa—" The judge seemed astounded. "But who would there be here to remember something that happened twenty years ago? And who, in any case, would want to make a gift of such a rare and expensive item to a complete stranger?"

The answer when it came was so low that I had difficulty making it out:

"There must have been *others,* Jason! Those people in Stillwater weren't the only ones who called Him master. Those worshipers of His—they still exist—they must! I believe it was one of them, carrying out his master's orders. As for where it came from in the first place, why, where else but—"

"No, Lucille, that's quite impossible," the judge cut her off. "Something I really can't allow myself to believe. If such things could be—"

"A madness the world could not face?"

"Yes, exactly!"

"Sam used to say the same thing. Nonetheless he sought the horror out, and brought me here with him, and then—"

"Yes, Lucille, I know what you believe happened then, but—"

"No buts, Jason—I want my son back. Help me, if you will, or don't help me. It makes no difference. I'm determined to find him, and I'll find him here, somewhere, I know it. If I have to, then I'll search him out alone, by myself, before it's too late!" Her voice had risen again, hysterically.

"No, there's no need for that," the old man cut in placatingly. "First thing tomorrow I'll find someone to help you. And we can get the Mounties from Nelson in on the job, too. They have a winter camp at Fir Valley only a few miles out of Navissa. I'll be able to get them on the telephone first thing in the morning. I'll need to, for the telephone will probably go out with the first bad snow."

"And you'll definitely find someone to help me person-ally—someone trustworthy?"

"That's my word. In fact I already know of one young man who might be willing. Of a very good family—and he's staying with me right now. You can meet him tomorrow—"

At this point I heard the scrape of chairs and pictured the two rising to their feet. Suddenly ashamed of myself to be standing there eavesdropping, I quickly returned to the library and pulled the door shut behind me. After some little time, during which the lady departed, I went again to Judge Andrews's study, this time tapping at the shut door and entering at his word. I found the old man worriedly pacing the floor.

He stopped pacing as I entered. "Ah, David. Sit down, please, there's something I would like to ask you." He seated himself, shuffling awkwardly in his chair. "It's dif-ficult to know where to begin—"

"Begin with Samuel R. Bridgeman," I answered. "I've had time to read his books now. Frankly, I find myself very interested."

"But how did you know—"

Thinking back on my eavesdropping, I blushed a little as I answered, "I've just seen Mrs. Bridgeman leaving. I'm guessing that it's her husband, or perhaps the lady herself, you want to talk to me about."

He nodded, picking up from his desk a golden medallion some two inches across its face, fingering its bas-relief work before answering. "Yes, you're right, but—"

"Yes?"

He sighed heavily in answer, then said, "Ah, well, I sup-pose I'll have to tell you the whole story, or what I know of it—that's the least I can do if I'm to expect your help." He shook his head. "That poor, demented woman!"

"Is she not quite . . . *right,* then?"

"Nothing like that at all," he answered hastily, gruffly.

"She's as sane as I am. It's just that she's a little, well, *disturbed.*"

He then told me the whole of the thing, a story that lasted well into the night. I reproduce here what I can remember of his words. They formed an almost unbroken narrative that I listened to in silence to its end, a narrative that only served to strengthen that resolution of mine to follow this mystery down to a workable conclusion.

"As you are aware," the judge began, "I was a friend of Sam Bridgeman's in our younger days. How this friendship came about is unimportant, but I also knew Lucille before they married, and that is why she now approaches me for help after all these years. It is pure coincidence that I live now in Navissa, so close to where Sam died.

"Even in those early days Sam was a bit of a rebel. Of the orthodox sciences, including anthropology and enthnology, few interested Sam in their accepted forms. Dead and mythological cities, lands with exotic names, and strange gods were ever his passion. I remember how he would sit and dream—of Atlantis and Mu, Ephiroth and Khurdisan, G'harne and lost Leng, R'lyeh and Theem'hdra, forgotten worlds of antique legend and myth—when by rights he should have been studying and working hard toward his future. And yet . . . that future came to nothing in the end.

"Twenty-six years ago he married Lucille, and because he was fairly well-to-do by then, having inherited a sizable fortune, he was able to escape a working life as we know it to turn his full attention to those ideas and ideals most dear to him. In writing his books, particularly his last book, he alienated himself utterly from colleagues and acknowledged authorities alike in those specific sciences upon which he lavished his 'imagination.' That was how they saw his—fantasies?—as the product of a wild imagination set free to wreak havoc among all established orders, scientific and theological included.

"Eventually he became looked upon as a fool, a naïve clown who based his crazed arguments in Blavatsky, in the absurd theories of Scott-Elliot, in the insane epistles of Eibon, and the warped translations of Harold Hadley Copeland, rather than in prosaic but proven historians and scientists. . . .

"When exactly, or why, Sam became interested in the theogony of these northern parts—particularly in certain beliefs of the Indians and half-breeds, and in Eskimo legends of yet more northerly regions—I do not know, but in the end he himself began to *believe* them. He was especially interested in the legend of the snow- or wind-god, Ithaqua, variously called 'Wind-Walker,' 'Death-Walker,' 'Strider in the Star-Spaces,' and others, a being who supposedly walks in the freezing boreal winds and in the turbulent atmospheric currents of far northern lands and adjacent waters.

"As fortune—or misfortune—would have it, his decision to pay this region a visit coincided with problems of an internal nature in some few of the villages around here. There were strange undercurrents at work. Secret semireligious groups had moved into the area, in many cases apparently vagrant, here to witness and worship at a 'Great Coming!' Strange, certainly, but can you show me any single region of this Earth of ours that does not have its crackpot organizations, religious or otherwise? Mind you, there has always been a problem with that sort of thing here. . . .

"Well, a number of the members of these so-called esoteric groups were generally somewhat more intelligent than the average Indian, half-breed, or Eskimo; they mainly New Englanders, from such decadent Massachusetts towns as Arkham, Dunwich, and Innsmouth.

"The Mounties at Nelson saw no threat, however, for this sort of thing was common here; one might almost say that over the years there has been a surfeit of it. On this occasion it was believed that certain occurrences in and about Stillwater and Navissa had drawn these rather polyglot visitors, for five years earlier there had indeed occurred a very large number of peculiar and still unsolved disappearances, to say nothing of a handful of inexplicable deaths at the same time.

"I've done a little research myself into just what happened, though I'm still very uncertain, but conjecture aside, hard figures and facts are—surprising?—no, they are downright disturbing!

"For instance, the *entire population* of one town, Stillwater, vanished overnight! You need not take my word for it—research it for yourself. The newspapers were full of it.

"Well, now, add to a background like this a handful of tales concerning giant webbed footprints in the snow, stories of strange altars to forbidden gods in the woods, and a creature that comes on the wings of the winds to accept living sacrifices—and remember, please, that all such appear time and again in the history and legends of these parts—and you'll agree it's little wonder that the area has attracted so many weird types over the years.

"Not that I remember Sam Bridgeman as being a 'weird type,' you understand; but it was exactly this sort of thing that brought him here when, after five years of quiet, the cycle of hysterical superstition and strange worship was again at its height. That was how things stood when he arrived here, and he brought his wife with him. . . .

"The snow was already deep to the north when they came, but that did nothing at all to deter Sam; he was here to probe the old legends, and he would never be satisfied until he had done just that. He hired a pair of French-Canadian guides, swarthy characters of doubtful backgrounds, to take him and Lucille in search of . . . of what? Dreams and myths, fairy tales and ghost stories?

"They trekked north, and despite the uncouth looks of

the guides, Sam soon decided that his choice of these two
men had been a good one; they seemed to know the region
quite well. Indeed, they appeared to be somehow, well,
cowed out in the snows, different again from when Sam had
found them, drunk and fighting in a Navissa bar. But then
again, in all truth, he had had little choice but to hire these
two, for with the five-year cycle of strangeness at its peak
few of Navissa's regular inhabitants would have ventured
far from their homes. And indeed, when Sam asked his
guides why they seemed so nervous, they told him it was all
to do with 'the season.' Not, they explained, the winter
season, but that of the strange myth-cycle. Beyond that
they would say nothing, which only excited Sam's curiosity
all the more—particularly since he had noticed that their
restlessness grew apace the farther north they trekked.

"Then, one calm white night, with the tents pitched and
a bright wood fire kindled, one of the guides asked Sam just
what it was that he sought in the snow. Sam told him,
mentioning the stories of Ithaqua the Snow-Thing, but got
no further; for upon hearing the Wind-Walker's name spoken, the French-Canadian simply refused to listen anymore. Instead, he went off early to his tent where he was
soon overheard muttering and arguing in a frightened and
urgent voice with his companion. The next morning, when
Sam roused himself, he discovered to his horror that he and
his wife were alone, that the guides had run off and deserted
them! Not only this, but they had taken all the provisions
with them. The Bridgemans had only their tent, the clothing
they stood in, their sleeping bags, and personal effects. They
had not even a box of matches with which to light a fire.

"Still, their case did not appear to be completely hopeless. They had had fair weather so far, and they were only
three days and nights out from Navissa. But their trail had
been anything but a straight one, so that when they set
about making a return journey it was pure guesswork on

Sam's part the correct direction in which to head. He knew something of the stars, however; and when the cold night came down, he was able to say with some certainty that they headed south.

"And yet lonely and vulnerable though they now felt, they had been aware even on the first day that they were not truly alone. On occasion they had crossed strange tracks, freshly made by furtive figures that melted away into the firs or banks of snow whenever Sam called out to them across the wintry wastes. On the second morning, soon after setting out from their camp in the lee of tall pines, they came upon the bodies of their erstwhile guides; they had been horribly tortured and mutilated before dying. In the pockets of one of the bodies Sam found matches, and that night—though by now they knew the pangs of hunger—they at least had the warmth of a fire to comfort them. But ever in the flickering shadows, just outside the field of vision afforded by the leaping flames, there were those furtive figures, silent in the snow, watching and . . . waiting?

"They talked, Sam and Lucille, huddled together in the door of their tent before the warming fire, whispering of the dead guides and how and why those men had come to such terrible ends; and they shivered at the surrounding shadows and the shapes that shifted within them. This country, Sam reasoned, must indeed be the territory of Ithaqua the Wind-Walker. At times, when the influence of old rites and mysteries was strongest, then the snow-god's worshipers—the Indians, half-breeds, and perhaps others less obvious and from farther parts—would gather here to attend His ceremonies. To the outsider, the unbeliever, this entire area must be forbidden, taboo! The guides had been outsiders . . . Sam and Lucille were outsiders, too. . . .

"It must have been about this time that Lucille's nerves began to go, which would surely be understandable. The intense cold and the white wastes stretching out in all directions, broken only very infrequently by the boles and snow-laden branches of firs and pines—the hunger eating at her insides now—those half-seen figures lurking ever on the perimeter of her vision and consciousness—the terrible knowledge that what had happened to the guides could easily happen again—and the fact, no longer hidden by her husband, that she and Sam were—lost! Though they were making south, who could say that Navissa lay in their path, or even that they would ever have the strength to make it back to the town?

"Yes, I think that at that stage she must have become for the most part delirious, for certainly the things she 'remembers' as happening from that time onward were delusion inspired, despite their detail. And God knows that poor Sam must have been in a similar condition. At any rate, on the third night, unable to light a fire because the matches had somehow got damp, events took an even stranger turn.

"They had managed to pitch the tent, and Sam had gone inside to do whatever he could toward making it comfortable. Lucille, as the night came down more fully, was outside moving about to keep warm. She suddenly cried out to Sam that she could see distant fires at the four points of the compass. Then, in another moment, she screamed, and there came a rushing wind that filled the tent and brought an intense, instantaneous drop in temperature. Stiffly, and yet as quickly as he could, Sam stumbled out of the tent to find Lucille fallen to the snow. She could not tell him what had happened, could only mumble incoherently of 'something in the sky!'

". . . God only knows how they lived through that night. Lucille's recollections are blurred and indistinct; she believes now that she was in any case more dead than alive. Three days and nights in that terrible white waste, wholly without food and for the greater part of the time without even the warmth of a fire. But on the morning of the next day—

"Amazingly everything had changed for the better overnight. Apparently their fears—that if they did not first perish from exposure they would die at the hands of the unknown murderers of the two guides—had been unfounded. Perhaps, Sam conjectured, they had somehow managed to pass out of the forbidden territory; and now that they were no longer trespassers, as it were, they were eligible for whatever help Ithaqua's furtive worshipers could give them. Certainly that was the way things seemed to be, for in the snow beside their tent they found tinned soups, matches, a kerosene cooker similar to the one stolen by the unfortunate guides, a pile of branches, and finally, a cryptic note that said, simply: 'Navissa lies seven miles to the southeast.' It was as if Lucille's vision of the foregoing night had been an omen of good fortune, as if Ithaqua Himself had looked down and decided that the two lost and desperate human beings deserved another chance. . . .

"By midday, with hot soup inside them, warmed and rested, having slept the morning through beside a fire, they were ready to complete their return journey to Navissa—or so they thought!

"Shortly after they set out, a light storm sprang up through which they pressed on until they came to a range of low, pinecovered hills. Navissa, Sam reckoned, must lie just beyond the hills. Despite the strengthening storm and falling temperature, they decided to fight on while they had the strength for it, but no sooner had they started to climb than nature seemed to set all her elements against them. I have checked the records and that night was one of the worst this region had known in many years.

"It soon became obvious that they could not go on through the teeth of the storm but must wait it out. Just as Sam had made up his mind to pitch camp, they entered a wood of thick firs and pines; and since this made the going easier, they pressed on a little longer. Soon, however, the storm picked up to such an unprecedented pitch that they knew they must take shelter there and then. In these circumstances they came across that which seemed a veritable haven from the storm.

"At first, seen through the whipping trees and blinding snow, the thing looked like a huge squat cabin, but as they approached it they could see that it was in fact a great raised platform of sorts, sturdily built of logs. The snow, having drifted up deeply on three sides of this edifice, had given it the appearance of a flat-roofed cabin. The fourth side being free of snow, the whole formed a perfect shelter into which they crept out of the blast. There, beneath that huge log platform whose purpose they were too weary even to guess at, Sam lit the kerosene stove and warmed some soup. They felt cheered by the timely discovery of this refuge, and since after some hours the storm seemed in no way about to abate, they made down their sleeping bags and settled themselves in for the night. Both of them fell instantly asleep.

"And it was later that night that disaster struck. How, in what manner Sam died, must always remain a matter for conjecture; but I believe that Lucille saw him die, and the sight of it must have temporarily broken her already badly weakened nerves. Certainly the things that she *believes* she saw, and one thing in particular that she believes happened that night, never could have been. God forbid!

"That part of Lucille's story, anyway, is composed of fragmentary mental images hard to define and even harder to put into common words. She has spoken of beacon fires burning in the night, of a 'congregation at Ithaqua's altar,' of an evil, ancient Eskimo chant issuing from a hundred adulatory throats—and of that which *answered* that chant, drawn down from the skies by the call of its worshipers. . . .

"I will go into no details of what she 'remembers' except to repeat that Sam died, and that then, as I see it, his poor wife's tortured mind must finally have broken. It seems certain, though, that even after the . . . horror . . . she must have received help from someone; she could not possibly have covered even a handful of miles in her condition on foot and alone—and yet she was found *here*, near Navissa, by certain of the town's inhabitants.

"She was taken to a local doctor, who was frankly astounded that, frozen to the marrow as she was, she had not died of exposure in the wastes. It was a number of weeks before she was well enough to be told of Sam, how he had been found dead, a block of human ice out in the snows.

"And when she pressed them, then it came out about the condition of his body, how strangely torn and mangled it had been, as if ravaged by savage beasts, or as if it had fallen from a great height, or perhaps a combination of both. The official verdict was that he must have stumbled over some high cliff onto sharp rocks, and that his body had subsequently been dragged for some distance over the snow by wolves. This latter fitted with the fact that while his body showed all the signs of a great fall, there were no high places in the immediate vicinity. Why the wolves did not devour him remains unknown."

Thus ended the judge's narrative, and though I sat for some three minutes waiting for him to continue, he did not do so. In the end I said, "And she believes that her husband was killed by? . . ."

"That Ithaqua killed him?—Yes, and she believes in rather worse things, if you can imagine that." Hurriedly then he went on, giving me no opportunity to question his meaning.

"One or two other things: First, Lucille's temperature. It has never been quite normal since that time. She tells me medical men are astounded that her body temperature never rises above a level that would be death to anyone else. They say it must be a symptom of severe nervous disorders but are at a loss to reconcile this with her otherwise fairly normal physical condition. And finally this." He held out the medallion for my inspection.

"I want you to keep it for now. It was found on Sam's broken body; in fact it was clenched in his hand. Lucille got it with his other effects. She tells me there is—something strange about it. If any, well, *phenomena* really do attach to it, you should notice them. . . ."

I took the medallion and looked at it—at its loathsome bas-relief work, scenes of a battle between monstrous beings that only some genius artist in the throes of madness might conceive—before asking, "And is that all?"

"Yes, I think so—no, wait. There is something else, of course there is. Lucille's boy, Kirby. He . . . well, in many ways it seems he is like Sam: impetuous, with a love of strange and esoteric lore and legend, a wanderer at heart, I suspect; but his mother has always kept him down, Earthbound. At any rate, he's now run off. Lucille believes that he's come north. She thinks perhaps that he intends to visit those regions where his father died. Don't ask me why; I think Kirby must be something of a neurotic where his father is concerned. This may well have come down to him from his mother.

"Anyway, she intends to follow and find him and take him home again away from here. Of course, if no evidence comes to light to show him positively to be in these parts, then there will be nothing for you to do. But if he really is here somewhere, then it would be a great personal favor to me if you would go with Lucille and look after her when she decides to search him out. Goodness only knows how it might affect her to go again into the snows, with so many bad memories."

"I'll certainly do as you ask, judge, and gladly," I answered immediately. "Frankly, the more I learn of Bridge-man, the more the mystery fascinates me. There *is* a mystery, you would agree, despite all rationalizations?"

"A mystery?" He pondered my question. "The snows are strange, David, and too much snow and privation can bring fantastic illusions—like the mirages of the desert. In the snow, men may dream while yet awake. And there again, there is that weird five-year cycle of strangeness that definitely affects this region. Myself, I suspect that it all has some quite simple explanation. A mystery?—I say the world is full of mysteries. . . ."

## III

That night I experienced my first taste of the weird, the inexplicable, the outré. And that night I further learned that I, too, must be susceptible to the five-year cycle of strangeness; either that, or I had eaten too well before taking to my bed!

There was first the dream of cyclopean submarine cities of mad angles and proportions, which melted into vague but frightful glimpses of the spaces between the stars, through which I seemed to walk or float at speeds many times that of light. Nebulae floated by like bubbles in wine, and strange constellations expanded before me and dwindled in my wake as I passed through them. This floating, or walking, was accompanied by the sounds of a tremendous striding, like the world-shaking footsteps of some ponderous giant, and there was (of all things) an ether wind that blew about me the scent of stars and shards of shattered planets.

Finally all of these impressions faded to a nothingness, and I was as a mote lost in the darkness of dead eons. Then there came another wind—not the wind that carried the odor of outer immensities or the pollen of blossoming planets—a tangible, shrieking gale wind that whirled me about and around until I was sick and dizzy and in dread of being dashed to pieces. And I awoke.

I awoke and thought I knew why I had dreamed such a strange dream, a nightmare totally outside anything I had previously known. For out in the night it raged and blew, a storm that filled my room with its roaring until I could almost feel the tiles being lifted from the roof above.

I got out of bed and went to the window, drawing the curtains cautiously and looking out—before stumbling back with my eyes popping and my mouth agape in an exclamation of utter amazement and disbelief. *Outside, the night was as calm as any I ever saw, with the stars gleaming clear and bright and not even a breeze to stir the small firs in the judge's garden!*

As I recoiled—amidst the rush and roar of winds that seemed to have their origin in my very room, even though I could feel no motion of the air and while nothing visibly stirred—I knocked down the golden medallion from where I had left it upon my window ledge. On the instant, as the dull yellow thing clattered to the smooth pine floor, the roaring of the wind was cut off, leaving a silence that made my head spin with its suddenness. The cacophony of mad winds had not "died away"—quite literally it had been *cut off!*

Shakily I bent to pick the medallion up, noticing that despite the warmth of my room it bore a chill that must have been near to freezing. On impulse I put the thing to my ear. It seemed that just for a second, receding, I could hear as in a sounding shell the rush and roar and hum of winds far, far away, winds blowing beyond the rim of the world!

In the morning, of course, I realized that it had all been a dream, not merely the fantastic submarine and interspatial sequences but also those occurrences following immediately upon my "awakening." Nevertheless, I questioned the judge as to whether he had heard anything odd during the night. He had not, and I was strangely relieved. . . .

Three days later, when it was beginning to look like Lucille Bridgeman's suspicions regarding her son were without basis—this despite all her efforts, and the judge's, to prove the positive presence of Kirby Bridgeman in the vicinity of Navissa—then came word from the Mounties at Fir Valley that a young man answering Kirby's description had indeed been seen. He had been with a mixed crowd of seemingly destitute outsiders and local layabouts camping in crumbling Stillwater. Observers—two aging but inveterate gold-grubbers, out on their last prospecting trip of the year before the bad weather set in—had mentioned seeing him. Though these gnarled prospectors had by no means been made welcome in Stillwater, nonetheless they had noted that this particular young man had appeared to be in a sort of trance or daze, and that the others with him had seemed to hold him in some kind of reverence; they had been tending to his needs and generally looking after him.

It was this description of the boy's condition (which made it sound rather as if he were not quite right in his head) that determined me to inquire tactfully of his mother about him as soon as the opportunity presented itself. For the last two days, though, I had been studying the handling and maintenance of a vehicle that the judge termed a "snow cat": a fairly large, motorized sledge of very modern design that he had hired for Mrs. Bridgeman from a friend of his in the town. The vehicle seemed a fairly economical affair, capable in suitable conditions of carrying two adults and provisions over snow at a speed of up to twenty miles per hour. It was capable, too, of a somewhat slower speed over more normal terrain. With such a vehicle two people might easily travel 150 miles without refueling, in comparative comfort at that, and over country no automobile could possibly challenge.

The next morning saw us setting out aboard the snow cat. Though we planned on returning to Navissa every second or third day to refuel, we had sufficient supplies aboard for at least a week. First we headed for Stillwater.

Following a fall of snow during the night, the track that led us to the ghost town was mainly buried beneath a white carpet almost a foot deep, but even so, it was plain that this barely fourth-class road (in places a mere trail) was in extremely poor repair. I recalled the judge telling me that very few people went to Stillwater now, following the strange affair of twenty years gone, and doubtless this accounted for the track's derelict appearance in those places where the wind had blown its surface clean.

In Stillwater we found a constable of the Mounties just preparing to leave the place for camp at Fir Valley. He had gone to the ghost town specifically to check out the story of the two old prospectors. Introducing himself as Constable McCauley, the Mountie showed us round the town.

Originally the place had been built of stout timbers, with stores and houses and one very ramshackle "saloon" bordering a main street and with lesser huts and habitations set back behind the street facades. Now, however, the main street was grown with grass and weeds beneath the snow, and even the stoutest buildings were quickly falling into dilapidation. The shacks and lesser houses to the rear leaned like old men with the weight of years, and rotten doorposts with their paint long flaked away sagged on every hand, threatening at any moment to collapse and bring down the edifices framing them into the snow. Here and there one or two windows remained, but warped and twisting frames had long since claimed by far the greater number, so that now sharp shards of glass stood up in broken rows from sills like grinning teeth in blackly leering mouths. A stained, tattered curtain flapped moldering threads in the chill midday breeze. Even though the day was fairly bright, there was a definite gloominess about Stillwater, an aura of something *not quite right,* of strange menace, seeming to brood like a mantle of evil about the place.

Overall, and ignoring the fact that twenty years had passed since last it knew habitation, the town seemed to be falling far too quickly into decay, almost as if some elder magic had blighted the place in an effort to return it to its origins. Saplings already stood tall through the snow in the main street; grass and weeds proliferated on window ledges, along facades, and in the black gaps where boards had fallen from the lower stories of the crumbling buildings.

Mrs. Bridgeman seemed to notice none of this, only that her son was no longer in the town . . . if he had ever been there.

In the largest standing building, a tavern that seemed to have fared better in its battle against decay than the rest of the town, we brewed coffee and heated soups. There, too we found signs of recent, if temporary, habitation, for the floor in one of the rooms was fairly littered with freshly empty cans and bottles. This debris, plus the blackened ashes of a fire built on stones in one corner, stood as plain testimony that the building had been used by that group of unknown persons whose presence the prospectors had reported.

The Mountie mentioned how chill the place was, and at his remark it dawned on me that indeed the tavern seemed colder inside (where by all rights it ought to have been at least marginally warmer) than out in the raw air of the derelict streets. I was about to voice this thought when Mrs. Bridgeman, suddenly paler by far than usual, put down her coffee and stood up from where she sat upon a rickety chair.

She looked first at me—a queer, piercing glance—then at McCauley. "My son was here," she abruptly said, as if she knew it quite definitely. "Kirby was here!"

The Mountie looked hard at her, then stared about the room in mystification. "There's some sign that your boy was here, Mrs. Bridgeman?"

She had turned away and for a moment did not answer. She seemed to be listening intently for something far off. "Can't you hear it?"

Constable McCauley looked at me out of the corner of his eye. He frowned. The room was very still. "Hear what, Mrs. Bridgeman? What is it?"

"Why, the wind!" she answered, her eyes clouded and distant. "The wind blowing way out between the worlds!"

Half an hour later we were ready to move again. The Mountie in the meantime had taken me to one side, to ask me if I didn't think the search we planned was just a little bit hazardous considering Mrs. Bridgeman's condition. Plainly he thought she was a bit touched. Perhaps she was! God knows, if what the judge told me was true, the poor woman had enough reason. Being ignorant of her real problem at that time, however, I shrugged her strangeness off, mentioning her relationship with her son as being obsessive out of all proportion to reality. In truth, this was the impression I had already half formed—but it did not explain the *other* thing.

I made no mention of it to the Mountie. For one thing, it was none of his business; and for another, I hardly wanted him thinking that perhaps I, too, was "a bit touched." It was simply this: in the derelict tavern—when Mrs. Bridgeman had asked, "Can't you hear it?"—I had in fact heard something. At the exact moment of her inquiry, I had put my hand into a pocket of my parka for a pack of cigarettes. My hand had come into contact with that strange golden medallion, and as my fingers closed upon the chill shape, I had felt a thrill as of weird energies, an electric tingle that seemed to energize all my senses simultaneously. I felt the cold of the spaces between the stars; I smelled again, as in my dreams, the scents of unknown worlds; for the merest fraction of a second there opened before me reeling vistas, incredible eons flashing by in a twinkling; and I, too, heard a wind—a howling *sentience* from far beyond the universe we know!

It had been so momentary, this—vision?—that I thought little more of it. Doubtless my mind, as I touched the medallion, had conjured in connection with the thing parts of that dream in which it had featured so strongly. That was the only explanation. . . .

I calculate that by 5:00 P.M. we must have been something like fifty miles directly north of Stillwater. It was there, in the lee of a low hill covered by tall conifers whose snow-laden branches bowed almost to the ground, that Mrs. Bridgeman called a halt for the night. Freezing, the snow already had a thin, crisp crust. I set up our two tiny bivouacs beneath a pine whose white branches formed in themselves something of a tent, and there I lit our stove and prepared a meal.

I had decided that it was time tactfully to approach Mrs. Bridgeman regarding those many facets of her story of which I was still ignorant; but then, as if there were not enough of mystery, I was witness to that which brought vividly back to me what the judge had told me of the widow's body temperature.

We had finished our meal, and I had prepared my bivouac for the night, spreading my sleeping bag and packing snow close to the lower outside walls of the tiny tent against freezing drafts. I offered to do the same for Mrs. Bridgeman, but she assured me that she could attend to that herself. For the moment she wanted "a breath of fresh air." That turn of phrase in itself might have been enough to puzzle me (the air could hardly have been fresher!) but in addition she then cast off her parka, standing only in sweater and slacks, before stepping out from under the lowered branches into the subzero temperatures of falling night.

Heavily wrapped, still I shivered as I watched her from the sanctuary of our hideaway beneath the tree. For half an hour she simply wandered to and fro over the snow, occasionally glancing at the sky and then again into the darkening distance. Finally, as I suddenly realized that I was quickly drawing close to freezing while waiting for her to come back to camp, I went stiffly out to her with her parka. She must by now, I believed, be very close to suffering from exposure. Blaming myself that I had not recognized sooner how terrifically cold it was, I came up to her and threw her parka about her shoulders. Imagine my astonishment when she turned with a questioning look, completely at ease and plainly quite comfortable, immensely surprised at my concern!

She must have seen immediately how cold I was. Chiding me that I had not taken greater care to keep warm, she hurried back with me to the bivouacs beneath the tree. There she quickly boiled water and made coffee. She drank none of the hot, reviving fluid herself, however, and I was so astounded at her apparent immunity to the cold that I forgot all about those questions I had intended to ask. Since Mrs. Bridgeman now plainly intended to retire and since my own sleeping bag lay warm and inviting inside my bivouac, I simply finished off the coffee, turned down the stove and lay down for the night.

I was suddenly tired, and the last thing I saw before sleeping was a patch of sky through the branches, illumined by brightly twinkling stars. Perhaps that picture of the heavens, imprinted upon my mind's eye as I fell asleep, colored my dreams. Certainly I dreamed of stars all night long, but they were uneasy dreams. The stars I saw were particularly sentient and paired like strange eyes; they glowed carmine against a moving black background of hideously suggestive design and immense proportions. . . .

In the morning over breakfast—cheese and tomato sandwiches, followed by coffee and fruit juice—I briefly mentioned Mrs. Bridgeman's apparent immunity to the cold, at which she looked at me with a very wry expression and said, "You may believe me, Mr. Lawton, when I tell you that I would give all of what little I have just once to feel the cold. It is this—*affliction*—of mine, an extremely rare condition that I contracted here in the north. And it has come out in—"

"In Kirby?" I hazarded the guess.

"Yes." She looked at me again, shrewdly this time. "How much did Judge Andrews tell you?"

I could not conceal my embarrassment. "He—he told me of your husband's death, and—"

"What did he say of my son?"

"Very little. He is not the kind of man to gossip idly, Mrs. Bridgeman, and—"

"And you suspect that there might be much to gossip about?" She was suddenly angry.

"I only know that I'm here, helping a woman look for her son, following her instincts and whims without question, as a favor to an old man. To be absolutely truthful, I suspect that there is a great mystery here; and I admit that I am addicted to mysteries, as curious as a cat. But my curiosity is without malice, you must believe that, and my only desire is to help you."

She turned away from me for a moment or two, and I thought she was still angry, but when she turned back her face was much more composed.

"And did the judge not warn you that there would be—danger?"

"Danger? Heavy snow is due, certainly—"

"No, the snow is nothing—I didn't mean the snow. The judge has Sam's books; have you read them?"

"Yes, but what danger can there be in mythology and folklore?" In fact, I guessed what she was getting at, but better to hear it from her own lips, as she "believed" it and as her husband had "believed" it before her.

"What danger in myths and legends, you ask?" She smiled mirthlessly. "I asked the same question of Sam when he wanted to leave me in Navissa. God, that I'd listened to him! What danger in folklore? I can't tell you directly—not without you thinking me a madwoman, as I'm sure the judge must more than half believe—but I'll tell you this: today we return to Navissa. On the way you can teach me how to drive the snow cat. I won't take you to horrors you can't conceive."

I tried to argue the point but she would say no more. We decamped in silence, packed the bivouacs and camp utensils aboard the cat, and then, despite a last effort on my part to dissuade her, she demanded that we head directly for Navissa.

For half an hour, traveling fairly slowly, we followed the course of a frozen stream between brooding fir forests whose dark interiors were made darker still by the shrouding snow that covered the upper branches. It was as I turned the snow cat away from the stream, around a smaller copse of trees to head more nearly south, that I accidentally came upon that which should have gone far toward substantiating Mrs. Bridgeman's hints of terrible dangers.

It was a large depression in the snow, to which I had to react quickly in order to avoid a spill, when we might easily have tumbled directly into it. I halted our machine, and we stepped down to take a closer look at this strangely sunken place in the snow.

Here the drift was deeper, perhaps three or four feet, but

in the center of the depression it had been compacted almost to the earth beneath, as if some great weight had rested there. The size of this concavity must have been almost twenty feet long by seven or eight feet wide, and its shape was something like—

Abruptly the judge's words came back to me—what he had mentioned of the various manifestations of Ithaqua, the Wind-Walker—*and particularly of giant, webbed footprints in the snow!*

But of course that was ridiculous. And yet . . .

I began to walk round the perimeter of the fantastic depression, only turning when I heard Mrs. Bridgeman cry out behind me. Paler than I had ever seen her before, now she leaned dizzily against the snow cat, her hand to her throat. I went quickly to her.

"Mrs. Bridgeman?"

"He—*He was here!*" she spoke in a horrified whisper.

"Your son?"

"No, not Kirby—*Him!*" She pointed, staring wide-eyed at the compacted snow of the depression. "Ithaqua, the Wind-Walker—that is His sign. And that means that I may already be too late!"

"Mrs. Bridgeman," I made a halfhearted attempt to reason with her, "plainly this depression marks the spot where a number of animals rested during the night. The snow must have drifted about them, leaving this peculiar shape."

"There was no snow last night, Mr. Lawton," she answered, more composed now, "but in any case your explanation is quite impossible. Why, if there had been a number of animals here, surely they would have left tracks in the snow when they moved. Look about you. There are no tracks here! No, this is the footprint of the fiend. The horror was here—and somewhere, at this very moment, my son is trying to search Him out, helped on by those poor devils that worship Him!"

I saw my chance then to avoid an early return to Navissa. If we went back now, I might never learn the whole story, and I would never be able to face the judge, having let him down. "Mrs. Bridgeman, it's plain that if we go south now we're only wasting time. I for one am willing to face whatever danger there may be, though I still can see no such danger. However, if some peril does face Kirby, then we won't be helping him any by returning to Navissa. It would help, though, if I knew the background story. Some of it I know already, but there must be a lot you can tell me. Now listen, we have enough fuel for about 120 miles more. This is my proposition: that we carry on looking for your son to the north. If we have not found him by the time our fuel reserves are halved, then we head back in a direct line for Navissa. Furthermore, I swear here and now that I'll never divulge anything you may tell me or anything I may see while you live. Now, then—we're wasting time. What do you say?"

She hesitated, turning my proposition over in her mind, and as she did so, I saw to the north the spreading of a cloud sheet across the sky and sensed that peculiar change of atmosphere that ever precedes bad weather. Again I prompted her: "The sky is growing more sullen all the time. We're in for plenty of snow—probably tonight. We really can't afford to waste time if we want to find Kirby before the worst of the weather sets in. Soon the glass will begin to fall, and—"

"The cold won't bother Kirby, Mr. Lawton—but you're right, there's no time to waste. From now on our breaks must be shorter, and we must try to travel faster. Later today I'll tell you what I can of . . . of everything. Believe what you will, it makes little difference, but for the last time I warn you—if we find Kirby, then in all probability we shall also find the utmost horror!"

## IV

With regard to the weather, I was right. Having turned again to the north, skirting dense fir forests and crossing frozen streams and low hills, by 10:30 A.M. we were driving through fairly heavy snow. The glass was far down, though mercifully there was little wind. All this time—despite a certainty in my heart that there would be none—nevertheless, I found myself watching out for more of those strange and inexplicable hollows in the snow.

A dense copse where the upper branches interlaced, forming a dark umbrella to hold up a roof of snow, served us for a midday camp. There, while we prepared a hot meal and as we ate, Mrs. Bridgeman began to tell me about her son, about his remarkable childhood and his strange leanings as he grew into a man. Her first revelation, however, was the most fantastic, and plainly the judge had been quite right to suspect that the events of twenty years gone had turned her mind, at least as far as her son was concerned.

"Kirby," she started without preamble, "is not Sam's son. I love Kirby, naturally, but he is in no wise a child of love. He was born of the winds. No, don't interrupt me, I want no rationalizations.

"Can you understand me, Mr. Lawton? I suppose not. Indeed, at first I, too, thought that I was mad, that the whole thing had been a nightmare. I thought so right until the time—until Kirby was born. Then, as he grew up from a baby, I became less sure. Now I know that I was never mad. It was no nightmare that came to me here in the snow but a monstrous fact! And why not? Are not the oldest religions and legends known to man full of stories of gods lusting after the daughters of men? There *were* giants in the olden times, Mr. Lawton. There still are.

"Do you recall the Wendy-Smith expedition of '33? What do you suppose he found, that poor man, in the fastnesses of Africa? What prompted him to say these words, which I know by heart: 'There are fabulous legends of star-born creatures who inhabited this Earth many millions of years before Man appeared and who were still here, in certain black places, when he eventually evolved. They are, I am sure, to an extent here even now.'

"Wendy-Smith *was sure,* and so am I. In 1913 two monsters were born in Dunwich to a degenerate half-wit of a woman. They are both dead now, but there are still whispers in Dunwich of the affair, and of the father who is hinted to have been other than human. Oh, there are many examples of survivals from olden times, of beings and forces that have reached godlike proportions in the minds of men, and who is to deny that at least some of them could be real?

"And where Ithaqua is concerned—why!—there are elementals of the air mentioned in every mythology known to man. Rightly so, for even today, and other than this Ithaqua of the Snows, there are strange winds that blow madness and horror into the minds of men. I mean winds like the *Foehn,* the south wind of Alpine valleys. And what of the piping winds of subterranean caverns, like that of the Calabrian Caves, which has been known to leave stout cavers white-haired, babbling wrecks? What do we understand of such forces?

"Our human race is a colony of ants, Mr. Lawton, inhabiting an anthill at the edge of a limitless chasm called infinity. All things may happen in infinity, and who knows what might come out of it? What do we know of *the facts* of anything, in our little corner of a never-ending universe, in this transient revolution in the space-time continuum? Seeping down from the stars at the beginning of time there were giants—beings who walked or flew across the spaces between the worlds, inhabiting and using entire systems at their will—and some of them still remain. What would the race of man be to creatures such as these? I'll tell you—we are the plankton of the seas of space and time!

"But there, I'm going on a bit, away from the point. The facts are these: that before I came to Navissa with Sam, he had already been told that he was sterile, and that after I left—after that horror had killed my husband—well, then I was pregnant.

"Of course, at first I believed that the doctors were wrong, that Sam had not been sterile at all, and this seemed to be borne out when my baby was born just within eight months of Sam's death. Obviously, in the normal scale of reckoning, Kirby was conceived before we came to Navissa. And yet it was a difficult pregnancy, and as a newborn baby he was a weedy, strange little thing—frail and dreamy and far too quiet—so that even without knowing much of children I nevertheless found myself thinking of his birth as having been . . . premature!

"His feet were large even for a boy, and his toes were webbed with a pink stretching of skin that thickened and lengthened as he grew. Understand, please, that my boy was in no way a freak—not visibly. Many people have this webbing between their toes; some have it between their fingers too. In all other respects he seemed to be completely normal. Well, perhaps not completely . . .

"Long before he could walk, he was talking—baby talk, you know—but not to me. Always it was when he was alone in his cot, and always when there was a wind. He could hear the wind, and he used to talk to it. But that was nothing really remarkable; grown children often talk to invisible playmates, people and creatures that only they can see; except that I used to listen to Kirby, and sometimes—

"Sometimes I could swear that the winds talked back to him!

"You may laugh if you wish, Mr. Lawton, and I don't suppose I could blame you, but there always seemed to be a wind about our home, when everywhere else the air was still. . . .

"As Kirby grew older this didn't seem to happen so frequently, or perhaps I simply grew used to it, I really don't know. But when he should have been starting school, well, that was out of the question. He was such a dreamer, in no way slow or backward, you understand, but he constantly lived in a kind of dreamworld. And always—though he seemed later to have given up his strange conversations with drafts and breezes—he had this fascination with the wind.

"One summer night when he was seven, a wind came up that threatened to blow the very house down. It came from the sea, a north wind off the Gulf of Mexico—or perhaps it came from farther away than that, who can say? At any rate, I was frightened, as were most of the families in the area where we lived. Such was the fury of that demon wind, and it reminded me so of . . . of another wind I had known. Kirby sensed my fear. It was the strangest thing, but he threw open a window and he shouted. He shouted right into the teeth of that howling, banshee storm. Can you imagine that? A small child, teeth bared and hair streaming, shouting at a wind that might have lifted him right off the face of the Earth!

"And yet in another minute the worst of the storm was over, leaving Kirby scolding and snapping at the smaller gusts of air that yet remained, until the night was as still as any other summer night. . . .

"At ten he became interested in model airplanes, and one of his private tutors helped him and encouraged him to design and build his own. You see, he was far ahead of other children his own age. One of his models created a lot of excitement when it was shown at an exhibition of flying models at a local club. It had a very strange shape; its underside was all rippled and warped. It worked on a gliding principle of my son's own invention, having no motor but relying upon what Kirby called his 'rippled-air principle.' I remember he took it to the gliding club that day, and that the other members—children and adults alike— laughed at his model and said it couldn't possibly fly. Kirby flew it for them for an hour, and they all marveled while it seemingly defied gravity in a fantastic series of flights. Then, because they had laughed at him, he smashed the model down to its balsa wood and tissue paper components to strew them like confetti at the feet of the spectators. That was his pride working, even as a child. I wasn't there myself, but I'm told that a designer from one of the big model companies cried when Kirby destroyed his glider. . . .

"He loved kites, too—he always had a kite. He would sit for hours and simply watch his kite standing on the air at the end of its string.

"When he was thirteen he wanted binoculars so that he could study the birds in flight. Hawks were of particular interest to him—the way they hover, motionless except for the rapid beating of their wings. They, too, seem almost to walk on the wind.

"Then came the day when a more serious and worrying aspect of Kirby's fascination with the air and flight came to light. For a long time I had been worried about him, about his constant restlessness and moodiness and his ominous obsession.

"We were visiting Chichén Itzá, a trip I hoped would take Kirby's mind off other things. In fact the trip had a twofold purpose; the other was that I had been to Chichén Itzá before with Sam, and this was my way of remembering how it had been. Every now and then I would visit a place where we had been happy before . . . before his death.

"There were, however, a number of things I had not taken into account. There is often a wind playing among those ancient ruins, and the ruins themselves—with their aura of antiquity, their strange glyphs, their history of bloody worship and nighted gods—can be . . . disturbing.

"I had forgotten, too, that the Mayas had their own god of the air, Quetzalcoatl, the plumed serpent, and I suspect that this was almost my undoing.

"Kirby had been quiet and moody during the outward trip, and he stayed that way even after freshening up and while we began to explore the ancient buildings and temples. It was while I was admiring other ruins that Kirby climbed the high, hideously adorned Temple of the Warriors, with its facade of plumed serpents, their mouths fanged and tails rampant.

"He was seen to fall—or jump—by at least two dozen people, mainly Mexicans, but later they all told the same story: how the wind had seemed almost to buoy him up; how he had seemed to fall in slow motion; how he had uttered an eerie cry before stepping into space, like a call to strange gods for assistance. And after that terrible fall, onto ancient stone flags and from such a great height? . . .

"It was a miracle, people said, that Kirby was unhurt.

"Well, eventually I was able to convince the authorities at the site that Kirby must have fallen, and I was able to get him away before he came out of his faint. Oh, yes, he had fainted. A fall like that, and the only result a swoon!

"But though I had explained away the incident as best I could, I don't suppose I could ever have explained the look on Kirby's face as I carried him away—that smile of triumph or strange satisfaction.

"Now all this happened not long after his fourteenth birthday, at a time when here in the north the five-year cycle of so-called 'superstitious belief and mass hysteria' was once more at its height, just as it is now. So far as I was concerned, there was an undeniable connection.

"Since then—and I blame myself that I've only recently discovered this—Kirby has been a secret saver, hoarding away whatever money he could lay his hands on toward some future purpose or ambition; and now of course I know that this was his journey north. All his life, you see, he had followed the trail of his destiny, and I don't suppose that there was anything I could have done to change it.

"A short time ago something happened to clinch it, something that drew Kirby north like a magnet. Now—I don't know what the end will be, *but I must see it*—I must find out, one way or the other, once and for all. . . ."

## V

By 1:30 P.M. we were once again mobile, our vehicle driving through occasional flurries of snow, fortunately with a light tail wind to boost us on our way. And it was not long before we came upon signs that warned of the presence of others there in that white waste, fresh snowshoe tracks that crossed our path at a tangent and moved in the direction of low hills. We followed these tracks—apparently belonging to a group of at least three persons—until they converged with others atop one of the low bald hills. Here I halted the snow cat and dismounted, peering out at the wilderness around and discovering that from here, between flurries of snow, I could roughly make out the site of our last camp. It dawned on me at once that this would have been a wonderful vantage point from which to keep us under observation.

Then Mrs. Bridgeman tugged at the sleeve of my parka, pointing away to the north where finally I made out a group of black dots against the pure white background straggling toward a distant pine forest.

"We must follow them," she declared. "They will be members of His order, on their way to the ceremonies. Kirby may even be with them!" At the thought her voice took on a feverish excitement.

"Quickly—we mustn't lose them!"

But lose them we did.

By the time we reached that stretch of open ground where first Mrs. Bridgeman had spied the unknown group, its members had already disappeared into the darkness of the trees some hundreds of yards away. At the edge of the forest I again brought our vehicle to a halt, and though we might easily have followed the tracks through the trees—which was my not-so-delicate companion's immediate and instinctive desire—that would have meant abandoning the snow cat.

Instead, I argued that we should skirt the forest, find a vantage point on its northern fringe, and there await the emergence of whichever persons they were who chose to wander these wastes at the onset of winter. To this seemingly sound proposal Mrs. Bridgeman readily enough agreed, and within the hour we were hidden away in a cluster of pines beyond the forest proper. There we took turns to watch the fringe of the forest, and while I took first watch, Mrs. Bridgeman made a pot of coffee. We had only unpacked our stove, deeming it unwise to make ourselves too comfortable in case we should need to be on the move in a hurry.

After only twenty minutes at my post I would have been willing to swear that the sky had snowed itself out for the day. Indeed I made just such a comment to my pale companion when she brought me a cup of coffee. The leaden heavens had cleared—there was hardly a cloud in sight in the afternoon sky—and then, as if from nowhere, there came the wind!

Instantly the temperature dropped, and I felt the hairs in my nostrils stiffening and cracking with each sniff of icy air. The remaining half cup of coffee in my hand froze in a matter of seconds, and a rime of frost sprang up on my eyebrows. Heavily wrapped as I was, still I felt the cold striking through, and I drew back into the comparative shelter of the trees. In all my meteorological experience I have never known or heard of anything like it before. The storm that came with the wind and the cold, rising up in the space of the next half hour, took me totally by surprise.

Looking up, through gaps in the snow-laden branches, I could plainly see the angry boiling up of clouds into a strange mixture of cumulonimbus and nimbostratus, where only moments before there had been no clouds at all! If the sky had seemed leaden earlier in the day, now it positively glowered. The atmosphere pressed down with an almost tangible weight upon our heads.

And finally it snowed.

Mercifully, and despite the fact that all the symptoms warned of a tremendous storm to come, the wind remained only moderate, but by comparison the snow came down as if it had never snowed before. The *husshh* of settling snow was quite audible as the huge flakes fell in gust-driven, spiraling myriads to the ground.

Plainly my watch on the forest was no longer necessary, indeed impossible, for such was the curtain of falling snow that visibility was down to no more than a few feet. We were stuck, but surely no more so than that suspicious band of wanderers in the forest—members of "His order," as Mrs. Bridgeman would have it. We would have to wait the weather out, and so would they.

For the next two hours, until about 5:00 P.M., I busied myself making a windbreak of fallen branches and packed snow until even the moderate wind was shut out of our hideaway. Then I built a small fire in the center of this sheltered area close to the snow cat. Whatever happened, I did not want the works of that machine put out of order by freezing temperatures.

During all this time Mrs. Bridgeman simply sat and brooded, plainly unconcerned with the cold. She was frustrated, I imagined, by our inability to get on with the search. In the same period, busy as I was with my hands, nevertheless I was able to ponder much of what had passed, drawing what half-formed conclusions I could in the circumstances.

The truth of the matter was that there did seem to be too many coincidences here for comfort, and personally I had already experienced a number of things previously unknown to me or alien to my nature. I could no longer keep from my mind memories of that strange dream of mine; similarly the odd sensations I had felt on contact with or in proximity to the yellow medallion of gold and obscure alloys.

Then there was the simple, quite definite fact—bolstered both by the judge and the widow Bridgeman alike, and by McCauley the Mountie—that a freakish five-year cycle of strange excitement, morbid worship, and curious cult activity *did* actually exist in these parts. And dwelling on thoughts such as these, I found myself wondering once again just what had happened here twenty years gone, that its echoes should so involve me here and now.

Patently it had not been—could not possibly have been—as Mrs. Bridgeman "remembered" it. And yet, apart from her previous nervousness and one or two forgivable lapses under emotional stress since then, she had seemed to me to be as normal as most women. . . .

Or had she?

I found myself in two minds. What of this fantastic immunity of hers to subzero temperatures? Even now she sat there, peering out into the falling snow, pale and distant and impervious still to the frost that rimed her forehead and dusted her clothes, perfectly comfortable despite the fact that she had once again shed her heavy parka. No, I was wrong, and it amazed me that I had fooled myself for so long. There was very little about this woman that was normal. She had known—*something*. Some experience to set her both mentally and physically aside from mundane mankind.

But could that experience possibly have been the horror she "remembered"? Even then I could not quite bring myself to believe.

And yet . . . what of that shape we had stumbled across in the snow, that deep imprint as of a huge webbed foot? My mind flashed back to our first night out from Navissa, when I had dreamed of a colossal shape in the sky, a shape with carmine stars for eyes!

—But this was no good. Why!—here I was, nervous as a cat, starting at the slightest flurry of snow out there beyond the heavy branches. I laughed at my own fancies, albeit shakily, because just for a second as I had turned from the bright fire I had imagined that a shadow moved out in the snow, a furtive figure that shifted just beyond my periphery of vision.

"I saw you jump, Mr. Lawton," my companion suddenly spoke up. "Did you see something?"

"I don't think so," I briskly answered, my voice louder than necessary. "Just a shadow in the snow."

"He has been there for five minutes now. We are under observation!"

"What? You mean there's someone out there?"

"Yes, one of His worshipers, I imagine, sent by the others to see what we're up to. We're outsiders, you know. But I don't think they'll try to do us any harm. Kirby would never allow that."

She was right. Suddenly I saw him, limned darkly against the white background as the whirling snow flurried to one side. Eskimo or Indian, I could not tell which, but I believe his face was impassive. He was merely—watching.

\* \* \*

From that time on the storm strengthened, with the wind
building up to a steady blast that drove the snow through
the trees in an impenetrable icy wall. Behind my barrier of
branches and snow we were comfortable enough, for I had
extended the shelter until its wall lay open only in a narrow
gap to the south; the wind was from the north. The snow on
the outside of the shelter had long since formed a frozen
crust, so that no wind came through, and the ice-stiffened
branches of the surrounding trees gave protection from
above. My fire blazed and roared in subdued imitation of
the wind, for I had braved half a dozen brief excursions
beyond the shelter to bring back armfuls of fallen branches.
Their trimmed ends burning, Indian fashion, where they
met like the spokes of a wheel to form the center of the fire,
these branches now warmed our small enclosure and gave
it light. They had burned thus all through the afternoon and
into the night.

It was about 10:00 P.M., pitch-black beyond the wall of
the shelter and still snowing hard, when we became aware
of our second visitor; the first had silently left us some hours
earlier. Mrs. Bridgeman saw him first, grabbing my elbow
so that I started to my feet and turned toward the open end
of our sanctuary. There, framed in the firelight, white with
snow from head to foot, stood a man.

A white man, he came forward shaking the snow from his
clothes. He paused before the fire and tipped back the hood
of his fur jacket, then shed his gloves and held his hands out
to the flames. His eyebrows were black, meeting across his
nose. He was very tall. After a while, ignoring me, he turned
to Mrs. Bridgeman. He had a strong New England accent
when he said, "It is Kirby's wish that you go back to
Navissa. He does not want you to be hurt. He says you
should return now to Navissa—both of you—and that you
should then go home. He knows everything now. He knows
why he is here, and he wants to stay. His destiny is the glory

of the spaces between the worlds, the knowledge and mysteries of the Ancient Ones who were here before man, godship over the icy winds of Earth and space with his Lord and Master. You have had him for almost twenty years. Now he wants to be free."

I was on the point of questioning his authority and tone when Mrs. Bridgeman cut me short. "Free? What kind of freedom? To stay here in the ice; to wander the icy wastes until any attempt to return to the world of men would mean certain death? To learn the alien lore of monsters spawned in black pits beyond time and space?"

Her voice rose hysterically. "To know no woman's love but sate his lust with strangers, leaving them for dead and worse in a manner that *only his loathsome father could ever teach him?*"

The stranger lifted his hand in sudden anger. "You dare to speak of Him like—" I sprang between them, but it was immediately apparent that I was not needed.

The change in Mrs. Bridgeman was almost frightening. She had been near to hysterics only seconds ago; now her eyes blazed with anger in her white face, and she stood so straight and regal as to make our unknown visitor draw back, his raised arm falling quickly to his side.

"Do *I* dare?" Her voice was as chill as the wind. "I am Kirby's mother! Yes, I dare—but what *you* have dared! . . . You would raise your hand to me?"

"I . . . it was only . . . I was angry." The man stumbled over his words before finding his former composure. "But all this makes no difference. Stay if you wish; you will not be able to enter the area of the ceremonies, for there will be a watch out. If you did get by the watch unseen—then the result would be upon your own heads. On the other hand, if you go back now, I can promise you fair weather all the way to Navissa. But only if you go now, at once."

My white-faced companion frowned and turned away to stare at the dying fire.

No doubt believing that she was weakening, the stranger offered his final inducement: "Think, Mrs. Bridgeman, and think well. There can only be one conclusion, one end, if you stay here—for you have looked upon Ithaqua!"

She turned back to him, desperate questions spilling from her lips. "Must we go tonight? May I not see my son just once? Will he be—?"

"He will not be harmed." She was cut off. "His destiny is—*great!* Yes, you must go tonight; he does not wish to see you, and there is so little—" He paused, almost visibly biting his tongue, but it seemed that Mrs. Bridgeman had not noticed his gaffe. Plainly he had been about to say "there is so little time."

My companion sighed and her shoulders slumped. "If I agree—we will need fair weather. That can be . . . arranged?"

The visitor eagerly nodded (though to me the idea that he might somehow contrive to control the weather seemed utterly ridiculous) and answered, "From now until midnight, the snow will lessen, the winds will die away. After that—" He shrugged. "But you will be well away from here before then."

She nodded, apparently in defeat. "Then we'll go. We need only sufficient time to break camp. A few minutes. But—"

"No buts, Mrs. Bridgeman. There was a Mountie here. He did not want to go away either. Now—" Again he shrugged, the movement of his shoulders speaking volumes.

"McCauley!" I gasped.

"That was not the Mountie's name," he answered me, "but whoever he was, he too was looking for this lady's son." He was obviously talking about some other Mountie from Fir Valley camp, and I remembered McCauley having mentioned another policeman who set out to search the wastes at the same time as he himself had headed for Stillwater.

"What have you done to him, to this man?" I asked.

He ignored me and, pulling on his gloves, again addressed Mrs. Bridgeman: "I will wait until you go." He pulled the hood of his jacket over his head, then stepped back out into the snow.

The conversation, what little there had been, had completely astounded me. In fact my astonishment had grown apace with what I had heard. Quite apart from openly admitting to what could only be murder, our strange visitor had agreed with—indeed, if my ears had not deceived me, he had *confirmed*—the wildest possible nightmares, horrors that until now, so far as I was aware or concerned, had only manifested themselves in the works of Samuel Bridgeman and others who had worked the same vein before him, and in the disturbed imagination of his widow. Surely this must be the final, utmost proof positive of the effect of the morbid five-year cycle on the minds of men? Could it be anything else?

Finally I turned to the widow to ask, "Are we actually going back to Navissa, after all your efforts? And now, when we're so close?"

First glancing cautiously out into the falling snow, she hurriedly shook her head, putting a warning finger to her lips. No, it was as I suspected; her almost docile concurrence, following that blazing, regal display of defiance, had merely been a ruse. She in no way intended to desert her son, whether he wished it or not. "Quickly—let's get packed up," she whispered. "He was right. The ceremony is tonight, it must be, and we haven't much time."

# VI

From then on my mind was given little time to dwell on anything; I simply followed Mrs. Bridgeman's directions to the letter, questioning nothing. In any case it was obvious that her game must now be played to outwit the enemy (I had come to think of the strange worshipers as "the enemy"), not to defeat them physically or to talk them down. That was plainly out of the question. If indeed they had resorted to murder in order to do whatever they intended to do, they would surely not let a mere woman stop them now.

So it was that when we set off south aboard the snow cat, in a direction roughly that of Navissa, I knew that it would not be long before we were doubling back on our tracks. And sure enough, within the half hour, at about 11:00 P.M., as we came over a low hill in the then very light snow, there Mrs. Bridgeman ordered a wide swing to the west.

We held this westward course for ten more minutes, then turned sharply to our right flank, bringing the snow cat once again onto a northerly course. For a further twenty minutes we drove through the light snow, which, now that it had the slackening north wind behind it, stung a little on my face. Then, again at Mrs. Bridgeman's direction, we climbed a thinly wooded slope to fetch a halt at the top not twenty minutes distant from our starting point. At the speed we had traveled, and given that the enemy had no machine comparable to our snow cat, we could not possibly have been followed; and here, sheltered by the thin trees and the still lightly falling snow, we should be quite invisible to the enemy somewhere to our front.

Now, while we paused for a moment, I once more found questions forming in my mind for which I had no answers, and I had no sooner decided to voice them than my pale companion pointed suddenly out through the thin branches of the trees on the summit of the hill in the direction of a great black forested area some half mile to the north. It was that same forest into which the enemy had vanished earlier in the day when we had been trailing them. Now at its four cardinal points, up sprang great fires of leaping red flame; and now too, coming to us on the wings of the north wind, faint and uneven we heard massed voices raised in a chilling ritual—the Rites of Ithaqua:

> *"Iä! Iä!—Ithaqua! Ithaqua!*
> *Ai! Ai! Ai!—Ithaqua!*
> *Ce-fyak vulg-t'uhm—*
> *Ithaqua fhtagn!*
> *Ugh!—Iä! Iä!—Ai! Ai! Ai!"*

Again and again, repeatedly the wind carried that utterly alien chorus to our ears, and inside me it seemed suddenly that my blood froze. It was not only this abhorrent chanting with its guttural tones, but also the *precision* of the—singing?—and the obvious familiarity of the voices with the song. This was no blind, parrotlike repetition of obscure vocal forms but a combination of a hundred or more perfectly synchronized voices whose soul-rending interpretation of a hideous alien liturgy had transformed it into this present awesome cacophony—a cacophony whose horror might indeed breach the voids between the worlds! Suddenly I knew that if there was an Ithaqua, then he must surely hear and answer the voices of his worshipers.

"Very little time now," my companion muttered, more to herself than to me. "The place of the ceremony must be central in that forest—and that's where Kirby is!"

I stared hard through the snow, which again was beginning to fall heavier, seeing that the nearest and most southerly of the four fires blazed some distance to the northeast of our position. The westerly fire was about a half a mile southwest of us.

"If we head directly between those two fires," I said, "entering the woods and heading straight for the most northerly fire, on the far side, then we should come pretty close to the center of the forest. We can take the snow cat to the edge of the trees, but from there we must go on foot. If we can grab Kirby and make a run for it—well, perhaps the cat can take three, at a push."

"Yes," she answered, "it's worth a try. If the worse comes to the worst . . . then at least I'll know what the end of it was. . . ."

With that I started up the cat's motor again, thankful that the wind was in our favor and knowing that under cover of the continuous chanting we stood a fair chance of driving right to the edge of the forest without being heard.

As we headed out across the white expanse of snow to the forest's edge, I could see in the heavens the glow of the fires reflected from the base of towering, strangely roiling nimbostratus. I knew then, instinctively, that we were in for a storm to end all storms.

At the edge of the forest, undetected so far, we dismounted and left the snow cat hidden in the lower branches of a great pine, making our way on foot through the forest's dark depths.

The going was of necessity very slow, and of course we dared show no light, but having progressed only a few hundred yards, we found that we could see in the distance the fires of individual torches, and the chanting came much louder and clearer. If there were guards, then we must have passed them by without attracting attention. The chanting was tinged now with a certain hysteria, a frenzy that built steadily toward a crescendo, charging the frosty air with unseen and menacing energies.

Abruptly, we came to the perimeter of a great cleared

area where the trees had been cut down to be built into a huge platform in the center. All about this platform a mongrel congregation of fur- and parka-clad men and women stood, their faces showing ruddy and wild-eyed in the light of numerous torches. There were Eskimos, Indians, Negroes, and whites—people from backgrounds as varied as their colors and races—over one hundred and fifty of them at a guess.

The time by then was rapidly approaching midnight, and the deafening, dreadful chanting had now reached such an intensity as to make any increase seem almost impossible. Nevertheless there *was* an increase, at which, with one final convulsive shriek, the entire crowd about the pyramidal platform prostrated themselves facedown in the snow—all bar one!

"Kirby!" I heard Mrs. Bridgeman gasp, as that one upright man, proud and straight backed, naked except for his trousers, commenced a slow and measured climb up the log steps of the platform.

*"Kirby!"* She shouted his name this time, starting forward and avoiding the arms I held out to restrain her.

*"He comes! He comes!"* The cry went out in a hiss of rapture from one hundred and fifty throats—drowning Lucille Bridgeman's shout—and suddenly I felt the expectancy in the air.

The prostrate figures were silent now, waiting; the slight wind had disappeared; the snow no longer fell. Only Mrs. Bridgeman's running figure disturbed the stillness, that and the flickering of torches where they stood up from the snow; only her feet on the ice-crusted surface broke the silence.

Kirby had reached the top of the pyramid, and his mother was running between the outermost of the encircling, prostrate figures when it happened. She stopped suddenly and cast a terrified glance at the night sky, then lifted a hand to her open mouth. I, too, looked up, craning my neck to see—and something moved high in the roiling clouds!

*"He comes! He comes!"* The vast sigh went up again.

Many things happened then, all in the space of a few seconds, comprising a total and a culmination beyond belief. And still I pray that what I heard and saw at that time, that everything I experienced, was an illusion engendered of too great a proximity to the mass lunacy of those who obey the call of the five-year cycle.

How best to describe it?

I remember running forward a few paces, into the clearing proper, before my eyes followed Mrs. Bridgeman's gaze to the boiling heavens where at first I saw nothing but the madly whirling clouds. I recall, however, a picture in my memory of the man called Kirby standing wide legged atop the great pyramid of logs, his arms and hands reaching in a gesture of expectancy or welcome up and outward, his hair streaming in a wind which sprang up suddenly *from above* to blow slantingly down from the skies. And then there is the vision that burns even now in my mind's eye of a *darkness* that fell out of the clouds like a black meteorite, a darkness grotesquely shaped like a man with carmine stars for eyes in its bloated blot of a head, and my ears still ring to the pealing screams of mortal fear and loathing that went up in that same instant from the poor, paralyzed woman who now saw and recognized the horror from the skies.

The Beast-God came striding down the wind, descending more slowly now than at first but still speeding like some great bird of prey to Earth, its fantastic splayfooted strides carrying it as if down some giant, winding, invisible staircase straight to the waiting figure atop the pyramid, until the huge black head turned and, from high above the trees, the thing called the Wind-Walker saw the hysterically screaming woman where she stood amid the prostrate forms of its worshipers—saw and *knew* her!

In midair the Being came to an abrupt, impossible halt—and then the great carmine eyes grew larger still, and the blackly outlined arms lifted to the skies in what was clearly an attitude of rage! One monstrous hand reached to the rushing clouds, and through them, to emerge but a split second later and hurl something huge and round to Earth. Still Mrs. Bridgeman screamed—loud, clear, and horrifically—as the unerringly hurled thing smashed down upon her with a roar of tortured air, flattening her instantly to the frozen ground and splintering into a mad bomb burst of exploding shards of—ice!

The scene about the log pyramid at that hellish moment must have been chaos. I myself was thrown in the rush of pressured air back into the trees, but in the next moment when I looked out again upon the clearing, all I could see was . . . blood!

The ice-torn, mangled bodies of a wide segment of worshipers were still tumbling outward from the blasted area where Mrs. Bridgeman had stood—a number of bloodied bodies still fell, lazily almost, like red leaves through the howling air; logs were beginning to burst outward from the base of the pyramid where flying chunks of ice had crashed with the force of grenades.

Nor was Ithaqua finished!

It seemed almost as if I could read this horror's thoughts as it towered raging in the sky: *Were these not His worshipers?—and had they not betrayed their faith in this matter, which was to have been His first meeting with His son on Earth? Well, they would pay for this error, for allowing this Daughter of Man, the mother of His son, to interfere with the ceremony!*

In the space of a few more seconds huge balls of ice were flung to Earth like a scattering of hailstones—but with far more devastating effect. When the last of them had hurled its ice-knife shards far and wide about the clearing, the snow was red with spouting blood; the screams of the torn and dying rose even above the howling devil-wind that Ithaqua had brought with Him from the star-spaces. The trees bent outward now from the clearing with the fury of that fiendish storm, and logs snapped and popped like matchsticks from the base of the platform at the crimson clearing's center.

But a change had taken place in the attitude of the lone figure standing wild and windblown at the top of the tottering pyramid.

While the gigantic, anthropomorphic figure in the sky had raged and ravaged, raining down death and destruction in the form of ice-globes frozen in his hands and snatched down out of the heavens, so the man-god-child, now grown to strange adulthood, had watched from his vantage point above the clearing all that transpired. He had seen his mother ruthlessly crushed to a raw, red pulp; he had watched the demoniac destruction of many, perhaps all of those deluded followers of his monstrous father. Still, in a dazed bewilderment, he gazed down upon the awful aftermath in the clearing—and then he laid back his head and screamed in a composite agony of frustration, horror, despair, and rapidly waxing rage!

And in that monumental agony his hellish heritage told. For all the winds screamed with him, roaring, howling, shrieking in a circular chase about the platform that lifted logs and tossed them as twigs in a whirlpool round and about in an impossible spiraling whirl. Even the clouds above rushed and clashed the faster for Kirby's rage, until at last his Father knew the anger of His son for what it was—but did He understand?

Down through the sky the Wind-Walker came again,

striding on great webbed feet through the currents of crazed air, arms reaching as a father reaches for his son—

—And at last, battered and bruised as I was and half unconscious from the wind's screaming and buffeting, I saw that which proved to me beyond all else that I had indeed succumbed to the five-year cycle of legend-inspired lunacy and mass hysteria.

For as the Ancient One descended, so His son rose up to meet Him—Kirby, racing up the wind in surefooted bounds and leaps, roaring with a hurricane voice that tore the sky asunder and blasted the clouds back across the heavens in panic flight—Kirby, expanding, exploding outward until his outline, limned against the frightened sky, became as great as that of his alien Sire—Kirby, Son of Ithaqua, whose clawing hands now reached in a raging blood lust, whose snarling, bestial, darkening features demanded revenge!

For a moment, perhaps astounded, the Wind-Walker stood off—and there were two darkly towering figures in that tortured sky, two great heads in which twin pairs of carmine stars glared—and these figures rushed suddenly together in such a display of aerial fury that for a moment I could make out nothing of the battle but the flash of lightning and roar of thunder.

I shook my head and wiped the frost and frozen blood droplets from my forehead, and when next I dared look at the sky, I could see only the fleeing clouds racing madly away—the clouds and high, high above them, two dark dots that fought and tore and dwindled against a familiar but now leering background of stars and constellations. . . .

Almost twenty-four hours have passed. How I lived through the horrors of last night I shall never know; but I did, and physically unscathed, though I fear that my mind may be permanently damaged. If I attempt to rationalize the thing, then I can say that there was a storm of tremendous and devastating fury, during the course of which I lost my mind. I can say, too, that Mrs. Bridgeman is lost in the snow, even that she must now be dead despite her amazing invulnerability to the cold. But of the rest? . . .

And on the other hand, if I forego all rationalizations and listen only to the little winds whispering among themselves behind my flimsy shelter? . . . Can I deny my own senses?

I remember only snatches of what followed the terrible carnage and the onset of the aerial battle—my return to the snow cat and how that machine broke down less than half an hour later in a blinding snowstorm; my frozen, stumbling fight against great white drifts with various items of equipment dragging me down; my bruising fall into a frozen hole in the snow whose *outlines* sent me in a renewed frenzy of gibbering terror across the wastes—until, exhausted, I collapsed here between these sheltering trees. I remember knowing that if I remained still where I had fallen, then I must die; and I recall the slow agony of setting up my shelter, packing the walls solid, and lighting the stove. There is nothing more, however, until I awakened around noon.

The cold roused me. The stove had long since burned itself out, but empty soup cans told me that somehow I had managed to feed myself before giving in to my absolute fatigue. I opened the reservoir of fuel in the stove's body and fired it again, once more attending to my hunger before drying out and warming my clothes item by item. Then, fortified and almost warm, heartened by a slight rise in the outside temperature, I set about the strengthening of this, my last refuge; for I knew by then that this was as far as I could hope to go.

At about 4:00 P.M. the sky told me that soon it must

storm again, and it was then that I thought to search out the snow cat and fetch precious fuel for my stove. I almost lost myself when the snow began to fall again, but by 6:00 P.M. I was back in my shelter having recovered almost a gallon of fuel from the crippled cat. I had spent at least fifteen futile minutes trying to restart the vehicle, which still lies where I found it less than half a mile from my refuge. It was then, knowing that I could live only a few days more at the outside, that I began to write this record. This is no mere foreboding, this grimly leering doom from which there can be no escape. I have given it some thought: I am too far from Navissa to stand even the slightest chance of making it on foot. I have food and fuel for three days at the most. Here . . . I can live for a few days more, and perhaps someone will find me. Outside, in some futile attempt to reach Navissa in the coming storm . . . I might last a day or even two, but I could never hope to cover all those miles in the snow.

It is about four in the morning. My wristwatch has stopped and I can no longer tell the time accurately. The storm, which I mistakenly thought had passed me by some miles to the north, has started outside. It was the roaring of the wind that roused me. I must have fallen asleep at my writing about midnight.

This is strange: the wind howls and roars, but through an opening in the canvas I can see the snow falling *steadily* against the black of the night, not hurried and hustled by the wind! And my shelter is too steady; it does not tremble in the gale. What does this mean?

I have discovered the truth. I am betrayed by the golden medallion, which, when I discovered the howling thing still in my pocket, I hurled out into a drift. There it lies now, outside in the snow, shrieking and screaming with the eternal crying of the winds that roar between the worlds.

To leave my shelter now would be certain death. And to stay? . . .

I must be quick with this, for He has come! Called by the demon howling of the medallion, He is here. No illusion this, no figment of my imagination but hideous face. *He squats without, even now!*

I dare not look out into His great eyes; I do not know what I might see in those carmine depths. But I do know now how I will die. It will be quick.

All is silence now. The falling snow muffles all. The black thing waits outside like a huge hunched blot on the snow. The temperature falls, drops, plummets. I cannot get close enough to my stove. This is how I am to pass from the world of the living, in the icy tomb of my tent, for I have gazed upon Ithaqua!

It is the end . . . frost forms on my brow . . . my lips crack . . . my blood freezes . . . I cannot breathe the air . . . my fingers are as white as the snow . . . the cold . . .

### NAVISSA DAILY
### The Snows Claim a Fresh Victim!

Just before the Christmas season, bad news has come out of Fir Valley camp where members of the Royal Canadian North-West Mounted Police have winter residence. During the recent lull in the weather, Constables McCauley and Sterling have been out in the wastes north of Navissa searching for traces of their former companion, Constable Jeffrey, who disappeared on routine investigations in October. The Mounties found no trace of Constable Jeffrey, but they did discover the body of Mr. David Lawton, an American meteorologist, who also disappeared in the snow in October. Mr. Lawton, accompanied by a Mrs. Lucille

Bridgeman, still missing, set out at that time in search of one Kirby Bridgeman, the woman's son. It was believed that this young man had gone into the wastes with a party of Eskimos and Indians, though no trace of this party has since been found. The recovery of Mr. Lawton's body will have to wait until the spring thaw; Constables McCauley and Sterling report that the body is frozen in a great block of clear ice that also encloses a canvas shelter and bivouac. The detailed report mentions that the eyes of the corpse are open and staring, as though the freezing took place with great rapidity. . . .

### NELSON RECORDER
**A Christmas Horror!**

Carol singers in the High Hill quarter of Nelson were astounded and horrified when, at 11:00 P.M., the frozen body of a young man crashed out of the upper branches of a tree in the grounds of No. 10 Church Street where they were caroling. Such was the force of its fall that the icy, naked figure brought down many branches with it. At least two of the witnesses state that the horribly mauled and mangled youth—whose uncommonly large and strangely webbed feet may help to identify him—fell not out of the tree but through it, as from the sky! Investigations are continuing.